SpringerBriefs in Psychology

SpringerBriefs present concise summaries of cutting-edge research and practical applications across a wide spectrum of fields. Featuring compact volumes of 50 to 125 pages, the series covers a range of content from professional to academic. Typical topics might include:

- A timely report of state-of-the-art analytical techniques
- A bridge between new research results as published in journal articles and a contextual literature review
- A snapshot of a hot or emerging topic
- An in-depth case study or clinical example
- A presentation of core concepts that readers must understand to make independent contributions

SpringerBriefs in Psychology showcase emerging theory, empirical research, and practical application in a wide variety of topics in psychology and related fields. Briefs are characterized by fast, global electronic dissemination, standard publishing contracts, standardized manuscript preparation and formatting guidelines, and expedited production schedules.

More information about this series at https://link.springer.com/bookseries/10143

Alexander Moreira-Almeida • Marianna de
Abreu Costa • Humberto Schubert Coelho

Science of Life After Death

Springer

Alexander Moreira-Almeida
School of Medicine – NUPES (Research Center in Spirituality and Health)
Federal University of Juiz de Fora - UFJF
Juiz de Fora – MG, Brazil

Marianna de Abreu Costa
School of Medicine – NUPES (Research Center in Spirituality and Health), Federal University of Juiz de Fora - UFJF
Juiz de Fora - MG, Brazil

Humberto Schubert Coelho
Department of Philosophy
Federal University of Juiz de Fora - UFJF
Juiz de Fora – MG, Brazil

ISSN 2192-8363   ISSN 2192-8371 (electronic)
SpringerBriefs in Psychology
ISBN 978-3-031-06055-7   ISBN 978-3-031-06056-4 (eBook)
https://doi.org/10.1007/978-3-031-06056-4

© The Author(s), under exclusive license to Springer Nature Switzerland AG 2022

This work is subject to copyright. All rights are solely and exclusively licensed by the Publisher, whether the whole or part of the material is concerned, specifically the rights of translation, reprinting, reuse of illustrations, recitation, broadcasting, reproduction on microfilms or in any other physical way, and transmission or information storage and retrieval, electronic adaptation, computer software, or by similar or dissimilar methodology now known or hereafter developed.

The use of general descriptive names, registered names, trademarks, service marks, etc. in this publication does not imply, even in the absence of a specific statement, that such names are exempt from the relevant protective laws and regulations and therefore free for general use.

The publisher, the authors and the editors are safe to assume that the advice and information in this book are believed to be true and accurate at the date of publication. Neither the publisher nor the authors or the editors give a warranty, expressed or implied, with respect to the material contained herein or for any errors or omissions that may have been made. The publisher remains neutral with regard to jurisdictional claims in published maps and institutional affiliations.

This Springer imprint is published by the registered company Springer Nature Switzerland AG
The registered company address is: Gewerbestrasse 11, 6330 Cham, Switzerland

Foreword

Many scientists doubt or ignore the pervasive evidence for a spiritual level of reality beyond what is experienced with the physical senses. They ignore the possibility of life after death even though the available evidence has vast implications for the goals and values of human life. In this valuable book on the science of life after death, the authors examine the evidence about life after death in a concise and balanced manner. They provide a reasonable basis for a wide audience to inform themselves about what is already known and what remains uncertain about the nature of conscious life.

The possibilities of spiritual life after death and of continuity of consciousness independent of the activity of the human brain are too important to be ignored based on authority, opinion, and prejudice. How do the functions of human consciousness, such as a sense of identity, the continuity of consciousness, freedom of will, and creativity emerge? How and why do we value ideals that we hold sacred, or feel a personal connection with nature, other people, and the future (Hay, 2007; Huxley, 2015) if we are merely mortal and separate things that seek to maximize their selfish desires (Dawkins, 1996)? Why has evolution moved relentlessly in the direction of ever-greater complexity, consciousness, and coherence if the mechanisms of consciousness are fully determined by unconscious matter (Eccles, 1989; Huxley, 1959; Wright, 1964)? What is causing leaders of modern governments, corporations, and cultural institutions to persist in actions that create conditions that impair human awareness, including fear about planetary climate change, violence from militarization, and distrust from social inequity (Cloninger, 2013)? Is the potential of human life for wisdom and other virtues so inadequate that we need to enhance human beings using artificial intelligence and other technological innovations (Harari, 2017; More & Vita-More, 2013)? Alternatively, can we better realize our untapped human potential by means of self-actualization and humane living conditions (Huxley, 2015)?

The possibility of life after death is the most ancient and perhaps the most important of all mysteries about the fundamental nature of life and consciousness. Wonder about spiritual life after death is well documented by elaborate ceremonial burials of the dead by the earliest *Homo sapiens* (literally, wise man) (Tattersall, 2012). In fact, narrative art, science, and spirituality are aspects of human self-aware consciousness, which emerged together about 100,000 years ago when evolution became aware of itself in the form of wise human beings (Cloninger, 2009; Zwir et al., 2022). Spirituality is defined as the search for what is beyond human existence (Cloninger, 2007). Most people report self-transcendent experiences at least occasionally in which they feel they are an inseparable aspect of something beyond human existence that is infinite and sacred (Cloninger, 2004; Hay, 2007). It is now well-documented that healthy physical, emotional, social, and cognitive aspects of health are strongly promoted by an understanding of life that admits the reality of self-transcendent experiences and spiritual life after death (Moreira et al., 2021).

The doubters of the possibility of life after death are left with the need to acknowledge such phenomena, but they only regard them as fortuitous health-promoting illusions, an inadvertent quirk of genetic or cultural evolution that favored such phenomena in *Homo sapiens* by virtue of their promoting physical survival and reproductive success and/or more cooperative and trusting communities (Dennett, 2006; Harari, 2015). Yet, in fact, all living organisms are aspects of complex systems in which symbiosis and cooperation are more involved in healthy social and ecological functioning than competition and aggression (Weiss & Buchanan, 2009). Things that appear separate to our physical senses are not separate from anything else at the most fundamental level of reality (Bohm, 2002; Cloninger, 2004). For example, cells are only partially separated by semipermeable boundaries that are actively engaged in exchange of materials in and out of themselves. All bits of matter and living things are composed of quanta that are constantly arising from, and returning to, a universal field of energy and information (i.e., the Higgs field). Consequently, no strictly physical model in science can account for consciousness or life itself (Cloninger & Cloninger, 2022).

If life survives death and consciousness is not explained by the physical brain alone, then it would seem to be necessary and wise to admit a transcendent aspect to reality, as have many of our most creative philosophers and scientists throughout history, as is well documented in this book. However, much is unknown about how physical, mental, and spiritual processes arise in human consciousness. Progress in our gaining deeper understanding of the nature of life and consciousness is impaired by closed-minded ideology, although some progress is being made in several fields, as discussed in this book.

Useful tools have been developed for measurement and experimentation that facilitate rigorous investigation of human functioning associated with possible spiritual phenomena, such as self-transcendent experiences, simultaneous creative discoveries, the gifts of child prodigies (Cloninger, 2004), and by the even more startling phenomena of mediumship, near death experiences, and cases of the reincarnation type, as discussed in this book. However, much of this work to construct adequately expanded scientific paradigms of human consciousness is still in its preliminary phases.

Ultimately, questions about the possibility of consciousness independent of the brain or of life after death cannot be fully grasped objectively or understood by the human intellect (i.e., analytical mind) (Chalmers, 1996, 1998). The limit of the intellect is to recognize that it is finite and unable to apprehend what is infinite or transcendent (Pascal & Krailsheimer, 1995). Human self-aware consciousness is a subjective experience with distinctive creative capacities that emerged uniquely in *Homo sapiens* about 100,000 years ago: it involves immediate insight about theories, aesthetics, and values that is intuitive and holographic with a spatial and temporal context spanning past, present, and future in a way that is vivid, fluent, and tolerant of ambiguity (Zwir et al., 2021, 2022). In contrast, the human analytical mind emerged in *Homo ergaster* (literally, working man), an early ancestor of both *Homo neanderthalensis* and *Homo sapiens* who lived a little less than 2 million years ago (Zwir et al., 2021, 2022). The analytic mind proceeds from uncertain assumptions and observations in a hierarchical way to converge on specific conclusions and procedures for seeking goals for individual benefit, which has different molecular processes for intentional self-control than those found in modern human self-awareness. Furthermore, most experimental work on the neural mechanisms of consciousness is primarily focused on unconscious procedural learning of habits and skills, which has been carried out by non-primates who evolved more than 41 million years ago and who have only rudimentary components of modern human sapience. Unlike unconscious procedural learning and the materialistic focus of the analytical mind, the self-awareness system of modern humans is uniquely capable of awareness of continuity of identity, self-transcendence, and other spiritual phenomena. Unfortunately, the development and expression of spiritual awareness can be impaired by conditions, such as fear, violence, and inequity, which are prevalent in contemporary societies (Zwir et al., 2022).

It is useful to recall that the scientific method cannot prove what is impossible, a fact that is often forgotten by skeptics to their eventual chagrin (Marshall, 2008; Siegel, 2017). The scientific method can only demonstrate the conditions under which something is predictable. Accordingly, in order to fully understand the potential of modern human consciousness, it is necessary to cultivate self-awareness through all its possible stages of human self-actualization, as described by intuitive sages and mystics since the beginning of history (Cloninger, 2004). Only then is it possible to describe the conditions for various aspects of possible human states of consciousness.

Fortunately, much is known about the processes of self-actualization and spiritual development, and this is beginning to be translated into paradigms amenable to reliable scientific description and understanding. Science, art, and spirituality are aspects of human self-awareness; they are not naturally in opposition or incompatible with one another because they are expressions of human wonder about the mysteries of life, which cannot be reduced to physical or brain processes alone. We are more than matter, so learning to integrate our habits to be in accord with our goals and values is what promotes our health and gives us meaning and purpose (Cloninger & Cloninger, 2022).

This book may provide an enlightening antidote to the blind skepticism that distorts human common sense so often in the widespread shadows of contemporary society. A healthy and happy life is also a virtuous life because these aspects of our conscious being are inseparable. Human fulfillment cannot be derived from external material things, such as unrestrained consumption, wealth, or power. Nor does self-actualization proceed from beliefs based on external authority or opinion. What is beneficial is that each of us reflects with an open mind about what others have observed about the mysteries of life, and then each looks deeply within their own being to discover who they are really and what gives them lasting satisfaction and meaning. This interior journey is interesting and fruitful for those who wish to expand their consciousness beyond the limits of their physical senses and analytical mind.

Anthropedia Foundation
Saint Louis, MO, USA
Washington University School of Medicine
Saint Louis, MO, USA

C. Robert Cloninger

References

Bohm, D. (2002). *Wholeness and the implicate order*. Routledge. Publisher description. http://www.loc.gov/catdir/enhancements/fy0664/2002031676-d.html

Chalmers, D. J. (1996). *The conscious mind: In search of a fundamental theory*. Oxford University Press. http://www.loc.gov/catdir/enhancements/fy0636/95036036-d.html

Chalmers, D. J. (1998). The problems of consciousness. *Advance in Neurology, 77*, 7–16; discussion 16–18. http://www.ncbi.nlm.nih.gov/entrez/query.fcgi?cmd=Retrieve&db=PubMed&dopt=Citation&list_uids=9709814

Cloninger, C. R. (2004). *Feeling good: The science of well-being*. Oxford University Press. http://www.loc.gov/catdir/toc/fy045/2003190053.html

Cloninger, C. R. (2007). Spirituality and the science of feeling good. *Southern Medical Journal, 100*(7), 740–743. http://www.ncbi.nlm.nih.gov/entrez/query.fcgi?cmd=Retrieve&db=PubMed&dopt=Citation&list_uids=17639764

Cloninger, C. R. (2009). Evolution of human brain functions: the functional structure of human consciousness. *Australian and New Zealand Journal of Psychiatry, 43*(11), 994–1006. https://doi.org/10.3109/00048670903270506

Cloninger, C. R. (2013). What makes people healthy, happy, and fulfilled in the face of current world challenges? *Mens Sana Monograph, 11*(1), 16–24. https://doi.org/10.4103/0973-1229.109288

Cloninger, C. R., & Cloninger, K. M. (2022). Self-transcendence. In J. R. Peteet (Ed.), *The virtues in psychiatric practice* (pp. 205–230). Oxford University Press. https://doi.org/10.1093/med/9780197524480.001.0001

Dawkins, R. (1996). *The blind watchmaker: Why the evidence of evolution reveals a universe without design*. Norton. http://www.blacksunjournal.com/m/dawkins_blindwatchmaker_1996_full.pdf

Dennett, D. C. (2006). *Breaking the spell: Religion as a natural phenomenon*. Viking. http://www.loc.gov/catdir/enhancements/fy1109/2005042415-d.html

Eccles, J. C. (1989). *Evolution of the brain: Creation of the self*. Routledge. https://doi.org/10.4324/9780203976661

Harari, Y. N. (2015). *Sapiens: a brief history of humankind* (First U.S. edition. ed.). Harper. https://www.worldcat.org/title/sapiens-a-brief-history-of-humankind/oclc/1075632482

Harari, Y. N. (2017). *Homo Deus: A brief history of tomorrow* (First U.S. edition. ed.). Harper, an imprint of HarperCollins Publishers. http://worldcat.org/oclc/951507538

Hay, D. (2007). *Something there: the biology of the human spirit*. Templeton Foundation Press. http://www.loc.gov/catdir/toc/ecip074/2006036297.html

Huxley, J. (1959). Foreword. In P. T. d. Chardin (Ed.), *The phenomenon of man* (pp. 11–28). Harper & Row. https://repository.library.georgetown.edu/handle/10822/762196

Huxley, J. (2015). Transhumanism. *Ethics in Progress*, 6(1), 12–16. https://doi.org/10.14746/eip.2015.1.2

Marshall, M. (2008, April 3, 2008). 10 impossibilities conquered by science. *New Scientist*, (2650). https://www.newscientist.com/article/dn13556-10-impossibilities-conquered-by-science/

More, M., & Vita-More, N. (2013). *The transhumanist reader: classical and contemporary essays on the science, technology, and philosophy of the human future*. Wiley-Blackwell. https://www.worldcat.org/title/transhumanist-reader-classical-and-contemporary-essays-on-the-science-technology-and-philosophy-of-the-human-future/oclc/1033918289

Moreira, P. A. S., Inman, R. A., & Cloninger, C. R. (2021). Virtues in action are related to the integration of both temperament and character: Comparing the VIA classification of virtues and Cloninger's biopsychosocial model of personality. *Journal of Positive Psychology*. https://doi.org/10.1080/17439760.2021.1975158

Pascal, B., & Krailsheimer, A. J. (1995). *Pensées* (Rev. ed.). Penquin Books USA. http://www.loc.gov/catdir/enhancements/fy1207/96103500-t.html

Siegel, E. (2017, 11/22/2017). Scientific proof is a myth. *Forbes*. https://www.forbes.com/sites/startswithabang/2017/11/22/scientific-proof-is-a-myth/?sh=5804fbeb2fb1

Tattersall, I. (2012). *Masters of the Planet: The search for our human origins*. St. Martin's Press. https://www.google.com/search?tbm=bks&hl=en&q=ISBN+978-0230108752

Weiss, K. M., & Buchanan, A. V. (2009). *The mermaid's tale: Four billion years of cooperation in the making of living things*. Harvard University Press. http://www.loc.gov/catdir/toc/fy0904/2008037860.html

Wright, S. (1964). Biology and the philosophy of science. *Monist*, 48, 265–290. https://www.jstor.org/stable/27901549

Zwir, I., Arnedo, J., Del-Val, C., Pulkki-Råback, L., Konte, B., Yang, S. S., Romero-Zaliz, R., Hintsanen, M., Cloninger, K. M., Garcia, D., Svrakic, D. M., Lester, N., Rozsa, S., Mesa, A., Lyytikäinen, L.-P., Giegling, I., Kähönen, M., Hernandez-Cuervo, H., Seppälä, I., . . . Cloninger, C. R. (2021). Three genetic-environmental networks for human personality. *Molecular Psychiatry*, 26(8), 3858–3875. https://doi.org/10.1038/s41380-019-0579-x

Zwir, I., Del-Val, C., Hintsanen, M., Cloninger, K. M., Romero-Zaliz, R., Mesa, A., Arnedo, J., Salas, R., Poblete, G. F., Raitoharju, E., Raitakari, O., Keltikangas-Jarvinen, L., de Erausquin, G. A., Tattersall, I., Lehtimaki, T., & Cloninger, C. R. (2022). Evolution of genetic networks for human creativity. *Molecular Psychiatry*, 27(1), 354–376. https://doi.org/10.1038/s41380-021-01097-y

Acknowledgments

This book is in many ways a synthesis of more than two decades of challenging and stimulating research on science, philosophy, and spirituality. We are deeply grateful to all those relatives, friends, colleagues, and institutions who helped and supported us during our academic journey. We are fortunate enough to have so many of them that we would undoubtedly do terrible injustices if we tried to nominate each of them. However, we must explicitly express our gratitude to the Federal University of Juiz de Fora (UFJF), which provided the support and academic environment of rigorous quest and freedom that allowed us to pursue careers in such a crucial and challenging area. Also, to our very esteemed colleagues from NUPES—Research Center in Spirituality and Health, who have come along with us in this fascinating odyssey of scientific and philosophical exploration of spirituality and human nature. Last but not least are our loved relatives who have graciously been inexhaustible springs of love, understanding, and encouragement.

AMA: my parents, Hélio and Elizabeth; my wife, Angelica; and my children, Victoria, Caio, and Laura.

MC: my mother, Suzy; and my husband, Tiago.

HSC: my grandmother, Suely; my parents, Luiz and Lívia; my wife, Camila; and my children, Alice and Ignácio.

Precisely for this book, we would like to acknowledge the invaluable financial support provided by Bial Foundation, IHPV, and Clinica Jorge Jaber. We are also indebted to the friends and colleagues who generously but carefully read this book's preliminary drafts: Daniel Assisi, Denise Paraná, Fernando Colugnati, Homero Vallada, Marcel Souto Maior, Miguel Farias, and Saulo Araujo. Finally, to Judy Jones, Springer Senior Editor in Behavioral Sciences, her generosity, competence, and enthusiasm for the book are exemplary!

Contents

1	**Introduction**..	1
	References..	3
2	**The Idea of Survival of the Soul in the History of Religions and Philosophy**..	5
	References..	10
3	**Setting the Scene: Addressing the Main Arguments Against Survival Hypothesis**..................................	13
	Neuroscience "Proves" that the Brain Generates Mind	13
	Principle of Parsimony: We Should Explain Mind Solely on a Material Basis...	16
	There Is No Mechanism for How the Mind Would Influence the Brain..	17
	Science Has Proved Physicalism and Survival Implies Supernaturalism.	19
	Survival Implies Cartesian Dualism that Is Rejected by Learned People.	21
	References..	22
4	**What Would Constitute Evidence for Personal Survival After Death?**..	27
	References..	31
5	**The Best Available Evidence for Life After Death**	33
	Mediumship	33
	Mrs. Leonora Piper..	34
	Chico Xavier...	36
	Other Mediumship Studies.....................................	39
	Alternative Explanations......................................	41
	Near-Death and Out-of-Body Experiences...........................	42
	Veridical Perceptions	43
	Alternative Explanations......................................	45

	Reincarnation	46
	Cases of Reincarnation Type	48
	Alternative Explanations	52
	References	54
6	**The Weight of the Whole Body of Evidence for Life After Death**	61
	Hypotheses Alternative to Survival	63
	Fraud	63
	Chance, Cryptomnesia, Fabrications of the Unconscious Mind, and Other Conventional Sources	64
	Living Agent Psi (LAP)	64
	References	70
7	**Cultural Barriers to a Fair Examination of the Available Evidence for Survival**	73
	References	76
8	**Conclusion**	79
	References	80
Index		81

About the Authors

Alexander Moreira-Almeida, MD, PhD Alexander Moreira-Almeida is Associate Professor of Psychiatry and director of the Research Center in Spirituality and Health (NUPES), School of Medicine, Federal University of Juiz de Fora (UFJF), Brazil. He is chair of the Section on Spirituality of the Latin American Psychiatric Association (APAL) and former chair of the Sections on Religion, Spirituality, and Psychiatry of the World Psychiatric Association (WPA 2014–20) and the Brazilian Psychiatric Association (ABP 2014–21). He has worked on the scientific investigation of spiritual experiences for more than 25 years, authoring more than 170 peer-reviewed papers and book chapters. His primary emphasis in the last decade has been the investigation of evidence of consciousness' activity beyond the brain, especially for evidence of survival of human consciousness after death. Alexander is editor of the books *Exploring Frontiers of the Mind-Brain Relationship* (Springer, 2012) and *Spirituality and Mental Health Across Cultures* (OUP, 2021).

Marianna de Abreu Costa, MD, PhD Marianna de Abreu Costa is a psychiatrist and has completed a PhD in psychiatry and behavioral sciences in the postgraduate program of Universidade Federal do Rio Grande do Sul, UFRGS. She is member of the Psychiatry and Spirituality Department of the Psychiatry Association of Rio Grande do Sul, DPE – APRS (2014–) and member of the Section on Spirituality of the Brazilian Psychiatric Association, ABP, Brazil (2019–). Marianna is a collaborator in studies emphasizing the integration of spirituality and religiosity in psychotherapy, and in studies that assess anomalous experiences and the mind-brain relationship developed at NUPES-UFJF.

Humberto Schubert Coelho, PhD Humberto Schubert Coelho is Associate Professor of Metaphysics and Modern Philosophy in the Department of Philosophy and co-director of the Research Center in Spirituality and Health (NUPES) at the Federal University of Juiz de Fora, and holder of the chair Nr. 23 of the Brazilian Academy of Philosophy. He was visiting researcher at the Ian Ramsey Centre for

Science and Religion, Oxford University (2019–2020). Humberto's research is devoted to the metaphysical ground for meaning and purpose in life, which includes a series of proofs for the existence of God and the immortality of the soul, along with cultural analyses of the relationship between philosophy, religion, and science.

1
Introduction

Excluding China and Russia, at least 60% of the population of the ten most populous countries in the world believe in life after death (Gedeshi et al., 2020). Even in highly secularized and industrialized Western Europe, only a 40% minority believe that "When people die, that is the end; there is NO life after death" (Pew Research Center, 2018). There is even some evidence that higher levels of education may actually be associated with greater acceptance that we have a soul and/or there is survival after death. This has been reported in countries as diverse as France (Pew Research Center, 2018), Russia (Pew Research Center, 2014), and Brazil (Curcio & Moreira-Almeida, 2019). Evidence has also been found that belief in the afterlife may have a positive impact on mental health, such as less anxiety and fewer suicidal thoughts as well as greater well-being, life satisfaction, and belief in an equitable world (Bradshaw & Ellison, 2010; Ellison et al., 2009; Flannelly et al., 2006, 2012; McClain-Jacobson et al., 2004).

The question of the survival of human personality after the death of the body is one of the most pervasive among cultures and religions, relevant to the practical, existential, and ethical dimensions of life, and many philosophers argued that the great and universal interest in the "immortality of the soul" is due to the very disposition of human reason that would lead us to these considerations (Coelho, 2022). The importance of the question can hardly be ignored. It can have a huge impact on worldview, meaning of life, and the nature of the universe and human beings. Because of this, it can also have a big impact on ethics[1] and grief.

This persistent predominance of belief in the afterlife clearly shows that it is a myth that it has no place in our modern world. This misconception is likely one of the many fruits of the myth of secularization (Josephson, 2017; Stark, 1999).

---

[1] To mention just one powerful example, one of the greatest modern philosophers, Immanuel Kant, concluded that ethical thinking necessarily involves at least the postulation of the immortality of the soul, without which moral action in general would be rendered meaningless (Kant, 1838, p. 113; 122–123).

© The Author(s), under exclusive license to Springer Nature Switzerland AG 2022
A. Moreira-Almeida et al., *Science of Life After Death*, SpringerBriefs in Psychology, https://doi.org/10.1007/978-3-031-06056-4_1

However, the high prevalence of a belief does not mean that it is factually true. Probably, most people think that the belief in survival after death is only a religious matter, not amenable to empirical investigation. But this matter of belief and speculation was transformed into a scientific one in the nineteenth century.[2] Nevertheless, most people will be unaware of the evidence produced by over 150 years of systematic scientific research on survival that has involved some of the brightest scientific and philosophical minds of the period, such as the Nobel laureates Charles Richet, Pierre Curie and Marie Curie, J. J. Thomson, Henri Bergson, and Lord Rayleigh (Daher et al., 2017; Gauld, 1982; Stevenson, 1977). This was well put by Ian Stevenson, probably the most prolific survival researcher of the twentieth century:

> The question of whether man survives after death is surely one of the most important that he can ask about himself. (...) despite formidable difficulties, the question is amenable to empirical investigation. (Stevenson, 1977, p. 168)

The present book will present and discuss as comprehensively as possible, given the constraints on space, the best available empirical evidence for survival of human consciousness after permanent bodily death, making a strong case for a rational, scientifically based belief in survival. Nevertheless, it is important to keep in mind that, in science, no single piece of evidence is enough to "prove" a hypothesis; the "proof" comes from the convergence of a wide variety of good supporting evidence. For example, Darwin responded to a botanist who requested of him a clear proof for natural selection (Candolle, 1862):

> natural selection ... hardly admits of direct proof or evidence. It will be believed in only by those who think that it connects & partly explains several large classes of facts. (Darwin, 1862)[3]

In a similar vein, the most compelling and "best available evidence" for survival comes from the convergence of the diversified and robust body of evidence of the persistence after death of the memory and character that makes up our personal identity. This evidence derives from a wide variety of human experiences (e.g., apparitions, mediumship, near-death experiences, cases suggestive of reincarnation, etc.) that mutually reinforce each other because they point to the same conclusion: survival of consciousness. It is worth noting that all this evidence has always been highly prevalent in all cultures and ages, since the dawn of *Homo sapiens* (Bozzano, 1997), and has been replicated, under strictly controlled studies, by many of the most competent scientists of the last 150 years.

However, even in science, the simple presentation of evidence is not enough if there are misguided philosophical and ideological prejudices clouding a fair, rational analysis of the body of empirical evidence. For example, Max Planck (1950),

---

[2] Until the beginning of the nineteenth century, most scientists and philosophers not only believed in the immortality of the soul but also that such belief was rationally inevitable. Along with the more famous proofs of God's existence, a large number of thinkers also presented proofs of human immortality, making it one of the main topics of philosophical analysis (Coelho, 2022).

[3] In another letter he stated that "the belief in natural selection" is "grounded ... chiefly from this view connecting under an intelligible point of view a host of facts" (Darwin, 1862).

the founder of quantum physics, by observing scientific controversies, noted a "remarkable" fact:

> A new scientific truth does not triumph by convincing its opponents and making them see the light, but rather because its opponents eventually die, and a new generation grows up that is familiar with it. (Planck, 1950, p. 34)

A striking case of dogmatic denial of evidence is the reaction to the publication, in the American Psychological Association flagship journal, of a comprehensive review of research into the reality of psi phenomena such as telepathy (Cardeña, 2018). An "academic reaction" to it offered basically no counterargument but rather the claim of an a priori impediment to the existence of the evidence. They argued that "claims made by parapsychologists cannot be true" because they would contradict physicalism,[4] that they consider to be "well-understood scientific principles." They did not go to the trouble of analyzing to invalidate a single piece of evidence provided because "pigs cannot fly," so they would know in advance that the data "are necessarily flawed and result from weak methodology or improper data analyses" (Reber & Alcock, 2020, p. 391).

This is why, before presenting and discussing the scientific evidence for survival, we will briefly introduce, in Chaps. 2 and 3, a contextual overview of survival belief and refute frequent misguided historical and epistemological arguments against the notion of survival. If you don't have any a priori rejection of the survival hypothesis, you might prefer to go directly to Chap. 4 which introduces the discussion of evidence for survival.

In summary, we will show in three steps how compelling the evidence is for survival. First, we will address and deconstruct the most common inaccurate philosophical, ideological, and methodological assumptions against survival. Then, we will present the best evidence for survival provided by the most rigorous studies. Finally, we will show that the most compelling evidence for survival is the striking convergence of findings from hundreds of scientific studies performed by a great number of researchers in a wide variety of spiritual and anomalous experiences. We will argue that survival of consciousness is the simplest and most natural scientific explanation for the whole body of evidence and that cultural and psychological prejudices are often the major stumbling blocks to accepting survival as a fact of nature.

References

Bozzano, E. (1997). *Povos primitivos e manifestações supranormais*. Editora Jornalistica FE.
Bradshaw, M., & Ellison, C. G. (2010). Financial hardship and psychological distress: Exploring the buffering effects of religion. *Social Science & Medicine, 71*(1), 196–204. https://doi.org/10.1016/j.socscimed.2010.03.015

---

[4] The assertion that everything that exists is ultimately composed of physical forces or particles.

Candolle, A. de D.-13 M. 2021. (1862). Letter no. 3603. In *Darwin Correspondence Project*. https://www.darwinproject.ac.uk/letter/DCP-LETT-3603.xml

Cardeña, E. (2018). The experimental evidence for parapsychological phenomena: A review. *American Psychologist, 73*(5), 663–677. https://doi.org/10.1037/amp0000236

Coelho, H. S. (2022). *História da Liberdade Religiosa: da Reforma ao Iluminismo*. Vozes.

Curcio, C. S. S., & Moreira-Almeida, A. (2019). Who does believe in life after death? Brazilian data from clinical and non-clinical samples. *Journal of Religion and Health, 58*(4), 1217–1234. https://doi.org/10.1007/s10943-018-0723-y

Daher, J. C., Damiano, R. F., Lucchetti, A. L. G., Moreira-Almeida, A., & Lucchetti, G. (2017). Research on experiences related to the possibility of consciousness beyond the brain: A bibliometric analysis of global scientific output. *The Journal of Nervous and Mental Disease, 205*(1), 37–47. https://doi.org/10.1097/NMD.0000000000000625

Darwin, C. (1862). *Letter no. 3608*. Darwin correspondence project. https://www.darwinproject.ac.uk/letter/DCP-LETT-3608.xml

Ellison, C. G., Burdette, A. M., & Hill, T. D. (2009). Blessed assurance: Religion, anxiety, and tranquility among US adults. *Social Science Research, 38*(3), 656–667. https://doi.org/10.1016/j.ssresearch.2009.02.002

Flannelly, K. J., Koenig, H. G., Ellison, C. G., Galek, K., & Krause, N. (2006). Belief in life after death and mental health: Findings from a national survey. *The Journal of Nervous and Mental Disease, 194*(7), 524–529. https://doi.org/10.1097/01.nmd.0000224876.63035.23

Flannelly, K. J., Ellison, C. G., Galek, K., & Silton, N. R. (2012). Belief in life-after-death, beliefs about the world, and psychiatric symptoms. *Journal of Religion and Health, 51*(3), 651–662. https://doi.org/10.1007/s10943-012-9608-7

Gauld, A. (1982). *Mediumship and survival: A century of investigations*. Heinemann.

Gedeshi, I., Pachulia, M., Rotman, D., Kritzinger, S., Poghosyan, G., Fotev, G., Kolenović-Đapo, J., Balobana, S., Baloban, J., Rabušic, L., Frederiksen, M., Saar, E., Ketola, K., Pachulia, M., Bréchon, P., Wolf, C., Rosta, G., Voas, D., Rovati, G., … Balakireva, O. (2020). *Joint EVS/WVS 2017–2021 Dataset (Joint EVS/WVS)* (1.0.0). World Values Survey Association. https://doi.org/10.14281/18241.2

Josephson-Storm, J. A. (2017). *The myth of disenchantment: Magic, modernity, and the birth of the human sciences*. The University of Chicago Press. https://ebookcentral.proquest.com/lib/uchicago/detail.action?docID=4827749

Kant, I. (1838). *Sämmtliche Werke* (Karl Rosenkranz, Friedrich Wilhelm Schubert. Bd. 1–12.). Leopold Voss.

McClain-Jacobson, C., Rosenfeld, B., Kosinski, A., Pessin, H., Cimino, J. E., & Breitbart, W. (2004). Belief in an afterlife, spiritual well-being and end-of-life despair in patients with advanced cancer. *General Hospital Psychiatry, 26*(6), 484–486. https://doi.org/10.1016/j.genhosppsych.2004.08.002

Pew Research Center. (2014, February 10). Russians return to religion, but not to Church. *Pew Research Center's Religion & Public Life Project*. https://www.pewforum.org/2014/02/10/russians-return-to-religion-but-not-to-church/

Pew Research Center. (2018). *Being Christian in Western Europe*. https://www.pewforum.org/2018/05/29/being-christian-in-westerneurope/

Planck, M. (1950). *Scientific autobiography, and other papers*. Williams & Norgate.

Reber, A. S., & Alcock, J. E. (2020). Searching for the impossible: Parapsychology's elusive quest. *The American Psychologist, 75*(3), 391–399. https://doi.org/10.1037/amp0000486

Stark, R. (1999). Secularization, R.I.P. *Sociology of Religion, 60*(3), 249–273. https://doi.org/10.2307/3711936

Stevenson, I. (1977). Research into the evidence of man's survival after death: A historical and critical survey with a summary of recent developments. *The Journal of Nervous and Mental Disease, 165*(3), 152–170. https://doi.org/10.1097/00005053-197709000-00002

2
The Idea of Survival of the Soul in the History of Religions and Philosophy

The idea of the survival of consciousness is closely related to the concept of spirituality. We, in common with a long tradition of prominent scholars and scientists of religion and spirituality, have proposed that a *sacred transcendent* is at the core of the ultimate meaning of spirituality (Coelho, 2021; Moreira-Almeida & Bhugra, 2021). This contact with the transcendent aspect of reality is a key component in virtually all spiritual traditions throughout history and cultures. This transcendent is understood to be of the utmost importance, what ultimately matters and, as a result, is sacred (Otto, 1958).

Usually, the human philosophical understanding about this transcendent aspect of reality involves the expectation or the certainty that human beings survive bodily death, as proposed, for example, by Christians, Muslims, Hindus, Buddhists, Jews, Ancient Greco-Roman traditions, and native Africans and Americans. Recent, large, worldwide surveys have shown that belief in an afterlife remains universal, accepted by most of the world's population (Curcio & Moreira-Almeida, 2019).

One of the main challenges to the strict, rational appreciation of the subject is the dogmatic attachment to a very specific definition of the soul/mind/spirit and, consequently, to a very specific definition of the afterlife or the state of said soul/mind/spirit after death, for example, the supposition that souls or spirits have to be completely immaterial minds. While we lack both conclusive philosophical demonstrations and empirical evidence that the spiritual part of human beings is conditioned by nature to manifest in a very specific fashion, this is (oddly) the prevalent belief among scholars and religious people.

In order to avoid dogmatic positions, it is important to consider them from a dispassionate, skeptical perspective, giving voice to whichever facts are worth analyzing.

The idea of survival, frequently presented as the idea of immortality of the soul, has seemingly been omnipresent among spiritual leaders, sages, and philosophers since the beginning of recorded history. This universality comes with one main advantage and one main disadvantage: the idea is common enough to be considered a "natural" one and varied enough in the form of expression to make the term ambiguous.[1]

The lack of a clear concept, or the assumption that only one very specific concept is acceptable, will bias the conclusions. If a scientific and rational concept of survival is to be achieved, the point of departure must necessarily be flexible with regard to the concept of soul. As the naturalist and poet Goethe said, we have no means to tell which part(s) of our constitution may remain after the death of our bodies, or for how long this part of us constituted by spiritual nature may remain (Scholz, 1934, p. 20–21), but we know that our knowledge of nature suggests that this spiritual part of humans is independent of the body.[2] Claiming to know *with perfect certainty* which part(s) will remain – or that none will remain – is a dogmatic position about what can or cannot be evidence of the soul's survival.

Since the idea of an immortal soul is time immemorial, the first mentions of it coincide with the oldest surviving texts, as in the case of the Babylonian epic *Gilgamesh*, where the hero endures many adventures in pursuit of biological immortality only to understand that spiritual life transcends the physical one (Shushan, 2009). In order to understand ancient conceptions of survival (or reincarnation, communication with the deceased…),[3] we have to keep in mind that most of them were veiled in symbolism. Phallic tomb monuments, for example, were commonly used to suggest that a new life would emerge from the grave.

In addition to the apparently general presence in all religions, anecdotal references to contact with the deceased have also been very common. There are plenty of references about oracles and at least a couple of famous ones from Apollonius of Tiana and Pliny the Younger, who described an encounter with a ghost in a haunted house, but for the sake of dramatic depiction, we should mention that Suetonius tells us that Nero was chased by his mother after killing her,[4] and the early Christian apologist and philosopher Justin the Martyr, for his part, asserts that necromancy was empirical proof of the immortality of the soul (Ogden, 2001, p. 158).

Along with anecdotal references, belief in divinatory or symbolic dreams and visionary dreamlike states was so commonplace and widely accepted that we must

---

[1] Aware of the semantic difficulties, modern thinkers have preferred more general claims based on a minimalist conception of survival. This is the reason why McTaggart argued that we should avoid the term immortal soul and opt for a less metaphysically demanding term, such as "the survival of men" or the survival of persons (McTaggart, 1930).

[2] At least in the sense of the mitigated or strong dualistic accounts of mind in the philosophy of mind. More fundamental and principiological approaches to the methodological and metaphysical grounds of the mind-body discussion also argue that non-subjective (materialistic) concepts of mind are meaningless (Düsing, 1997).

[3] Rohde describes Greek funerary rituals – but the same is valid for virtually every culture – that include many procedures which are illogical from a materialistic perspective, such as crowning the dead with flowers, dressing them for glory, or equipping them for battle (Rohde, 2010, 219–224). And many cultures even developed ways to venerate the souls of the dead long after the funeral, or to guide the spirits to a new incarnation (Obst, 2009).

[4] The belief that spirits would torment criminals was so common that it justified the myth of the Furies.

wonder about the causes for the persuasiveness of such beliefs (Miller, 1994). This question gains substance if we consider that this belief in spiritual visitations and communication through dreams is still present in contemporary society and seems to ignore cultural boundaries.

Dreams are of radical importance for animist cultures, ancient and contemporary, often including contact with deceased relatives and friends, and veridical perceptions or premonitions suggestive of a spiritual reality. Sometimes, these dreams have more observable – nowadays, testable? – consequences, such as physical or mental healing. Among the historical references about healings and "miracles" in Antiquity, there are many mentions of out-of-body experiences or *instructions* received in dreams about procedures, drugs, or behavioral changes necessary for the cure (Weinreich, 1909). Of course, the idea that dreams may reveal something or that they may be a stage for encounters with the dead presupposes a spiritual reality.

From Egypt to India, from Babylon to Greece, in Zoroastrianism, Christianity, and Islam, moral consequences related to an afterlife seem to be almost as ubiquitous as the very belief in the afterlife, often identifying very distinct places for punishment and reward, as well as some kind of trial (Coelho, 2015; Rohde, 2010, p. 307).[5] Both artistic depictions and moral teachings refer to the deceased as "still living" in the afterlife.

While religions presuppose or imply spiritual experiences with an actual spiritual reality (Bozzano, 1997; Kripal, 2010; Lang, 1898), the greatest thinkers frequently preferred philosophical analysis of the possibility and implications of the concept of a surviving soul. Since the fundamental tools of rational thinking, such as logic and dialectics, were formulated by Plato and Aristotle, most scientific thinking, and the even broader scope of rational investigation and arts, flowered from the soil of their (rather spiritualistic) philosophy.

Although previous philosophers had presented arguments in favor of the survival of the soul, it was Plato who shaped the entire history of this discussion by connecting the subject with the ultimate grounds for rational thought: the mental or subjective constitution of ideas as regulative principles for all judgements. In his *Phaedo*, Plato argued that only an intellectual substance could interact with ideas, and since we only think by means of ideas/form, a part of us would clearly be intellectual/

---

[5] As a single paradigmatic example, we may consider the case of Zoroastrianism, which strongly influenced the moral dualism of (at least) Judaism, Christianity, Islam, and Manicheism. Zarathustra (westernized Zoroaster) was an Iranian sage that may have lived around 1100 to 2000 years B.C. Like pretty much every founder of a new religion, he was a reformer that criticized exterior rituals and materialism in favor of a more universal conception of one supreme God whose main characteristics were wisdom and goodness. In the afterlife, the destinies of the souls are separated after judgment. The exactness of the correspondence between moral principles and the respective destinies to which they lead suggests that one's state after death would be a direct consequence of one's moral nature and behavior (Nigosian, 1993, p. 23). Zoroastrian moral teaching is extremely libertarian and consequentialist (Nigosian, 1993, 23). The idea of impediments to perfect behavior has no place, leaving the rigorous consequences of voluntaristic optimism as the only possible explanation for every virtuous or evil deed. As in the copious examples throughout religions, freewill is strongly asserted and good is its own reward.

spiritual.[6] The argument is far more powerful than it seems, for the most common objection to it is the philosophically unsustainable notion that ideas and patterns of thought are simply concocted in the course of social history. Moreover, if the intellect is familiar with the eternal and a stranger to matter, it is likely, Plato concluded, that our mind/intellect is a spiritual substance and survives the dissolution of our body (Plato, 1913, p. 274 [78e-83c]).

Plato's approach to the problem of the soul, thus, is impossible to dissociate from his approach to the problems of reality and knowledge, for two things are necessary to give systematic integrity to our understanding of life and the world: (1) we have to consider that the raising of questions about the ultimate reality, truth, and knowledge is itself a movement that forces subjectivity into scientific and philosophical inquiry; (2) what we actually want with scientific and philosophical investigation is to find the truth, wisdom, and virtue, but such goals have some distinct features that demand an idealistic or spiritualistic solution. According to Plato, and basically every rationalist after him, including most mathematicians, for example, it would be irrational to consider materialistic solutions for the most fundamental questions.

Aristotle tried to capture the soul in its manifestations, both in terms of biology and ethics. In both cases, a difference can be observed between our more or less automatic inclinations and our sensorial reactions to the environment (including desires) and the free intelligent deliberations of our consciousness (Hicks, 2015, III.4, [427a6–427b26]), which do not seem to follow a physical order but rather a higher one.[7]

Plato and Aristotle shaped the structure of rational thinking, and both of them discussed the spiritual nature of humans only with rational arguments. In fact, most Western philosophers and scientists, from Pythagoras and Cicero to Newton and from G. Leibniz and Robert Boyle (Sommer, 2013) to Max Planck and Kurt Gödel (Wang, 1996), considered the immortality of the soul a *sequitur* of the very premises that formed the basis for scientific and logical thought. The fact that most of the greatest intellects of human history were strongly persuaded about the soul's survival – and many of them through the force of Plato's arguments – is an oft overlooked fact, and it calls for better exploration and understanding.

The transcendental philosophy of Immanuel Kant, one of the pinnacles of modern thought, represented for its followers the victory of spiritualism over materialism. In a seemingly complex relationship with the subject, Kant both mocked the Swedish medium Emanuel Swedenborg for offering a dreamy narrative about the spiritual world and offered himself elaborated arguments in favor of the immortality of the soul. Rooting the concept of an afterlife in his ethics, Kant argued that the fact that all cultures and individuals consider the possibility of survival is a consequence of the very structure of reasoning. It would be not a fortuitous occurrence, therefore, that all humans at least "entertain" the idea of an afterlife. This idea would be

---

[6] Additionally, Plato concluded not only that spiritual and material reality differ from each other but also that the immediate presence of intellectual knowledge in us suggests previous existences of our souls (Plato, 1913, p. 256 [74a-78a]).

[7] In his ethics and in his cosmology, Aristotle also argues that the best part of every being is divine.

inevitably produced by our rational analysis of the meaning and purpose of life (Kant, 2003, 107–123).

Existential arguments would grow in sophistication a century after Kant's death, and many thinkers would emphasize that *existential courage* and happiness have strong correlation with metaphysical worldviews that justify suffering and purpose in perspective of a larger cosmic order, which includes an expanded notion of life (James, 2002). Far from being purely abstract, philosophy also frequently proves itself to be the grounds and the cause for behavior, interpretation of life, and existential attitudes, since they are all directed by one's view of the world and life.

It is logically evident that Nihilism will have detrimental consequences to one's mood, resilience, and capacity to find meaning for life in moments of despair, while an optimistic or even neutral metaphysics – in the sense of agnostically admitting the *possibility* of ultimate meaning and purpose for human life – has been proved a predicting factor of a happier and healthier life (VanderWeele et al., 2022).

It is unlikely that someone with a nihilist philosophy would invest the same effort and possess the same disposition in a project as one persuaded that life has profound meaning and is worth living and that good deeds may be rewarded. Optimism and the basic motivational idea that we can be masters of our own destiny, concepts well familiar to the greatest souls, also presuppose some level of separation between mental and physical reality.

Johann G. Fichte, one of the keenest philosophers of all time, argued that science and materialism were in opposition to each other and that thought itself was evidence of a spiritual reality (Fichte, 1965). More persuasively than Descartes or Berkeley, Fichte demonstrates that our minds are the only immediate realities we can ever be aware of and that any feeling or thought about something else – such as material objects – is dependent on our minds. Even the atheist Schopenhauer agreed that philosophy could not be grounded on a non-mental substance such as the concept of matter and, therefore, that matter had to depend on a more spiritual substance.

While the metaphysical rejection of the concept of an afterlife seems inexcusable, other critics prefer more ambiguous terms to reject the transcendent elements of human thought as anti-scientific. One classical attempt of exclusion of this element is to situate some discussions outside the reach of rational or, at least, scientific inquiry. The very popular academic myth of conflict and mutual exclusion of science and religion is among the most successful tools in the war of narratives on this matter since the Enlightenment (Coelho, 2022).

F. Schleiermacher, the founder of philosophy of religion, argued that a scientific study of religion should start from the living experience of the faithful instead of a priori disqualification of the phenomena as some sort of psychological or cultural delusion. For the faithful, religion is not something that humans do or produce; it is something that they universally perceive as *being done to them*, being acted upon them, from the outside, and the religious felling is characterized by the consciousness of our dependency on the source of these occurrences (Sockness & Gräb, 2010). Now, we may well interpret this feeling as a natural tendency to detect vague agency upon our lives, as part of the contemporary cognitive science of religion does, but it would be unfair to ignore the very constitution of such a universal experience.

The historical developments of philosophy and the sciences in the last couple of centuries, however, have led to a state of distrust in the metaphysical arguments for the soul's immortality, and, although *never properly proved false*, these metaphysical arguments were ultimately rejected as insufficiently based on empirical reality (Davies, 2000). As we will discuss later, empirical research into survival has also been bitterly opposed, to the extent that neither philosophical nor empirical research have received much credit in the age of physicalism. Challenged in the last couple of centuries, the idea of a surviving soul was eventually depicted as unsophisticated in the academic culture (Josephson, 2017; Kripal, 2010; Sommer, 2013, 2016; Stark, 1999). Despite the materialistic culture or, more likely as a reaction to it, many scientists, philosophers and scholars since the nineteenth century have initiated a quest to investigate empirical evidence for survival.

References

Bozzano, E. (1997). *Povos primitivos e manifestações supranormais*. Editora Jornalistica FE.
Coelho, H. S. (2015). As exposições de François Decret e Pio de Luis sobre o maniqueísmo. *Revista Ética e Filosofia Política, 1*(18). https://doi.org/10.34019/2448-2137.2015.17659
Coelho, H. S. (2021). O que é e qual é o lugar do transcendente na filosofia? Uma nota a partir de G. W. F. Hegel. *Ética e Filosofia Política XXIV*.
Coelho, H. S. (2022). *História da Liberdade Religiosa: Da Reforma ao Iluminismo*. Vozes.
Curcio, C. S. S., & Moreira-Almeida, A. (2019). Who does believe in life after death? Brazilian data from clinical and non-clinical samples. *Journal of Religion and Health, 58*(4), 1217–1234. https://doi.org/10.1007/s10943-018-0723-y
Davies, B. (Ed.). (2000). *Philosophy of religion: A guide and anthology*. Oxford University Press.
Düsing, K. (1997). *Selbstbewusstseinsmodelle; Moderne Kritiken und systematische Entwürfe zur konkreten Subjektivität*. Brill/Fink.
Fichte, J. G. (1965). *Gesamtausgabe der Bayerischen Akademie der Wissenschaften. 1,2: Werke 1793–1795*. Frommann-holzboog Verlag.
Hicks, R. D. (2015). *Aristotle de Anima: With translation, introduction and notes*. Cambridge University Press.
James, W. (2002). *Varieties of religious experience: A study in human nature*. Routledge.
Josephson-Storm, J. A. (2017). *The myth of disenchantment: Magic, modernity, and the birth of the human sciences*. The University of Chicago Press.
Kant, I. (2003). *Kritik der praktischen Vernunft*. Felix Meiner Verlag.
Kripal, J. J. (2010). *Authors of the impossible: The paranormal and the sacred*. The University of Chicago Press.
Lang, A. (1898). *The making of religion*. Longmans, Green and Co.
McTaggart, J. M. E. (1930). *Some dogmas of religion*. Edward Arnold.
Miller, P. C. (1994). *Dreams in late antiquity: Studies in the imagination of a culture*. Princeton University Press.
Moreira-Almeida, A., & Bhugra, D. (2021). Religion, spirituality and mental health: Setting the scene. In A. Moreira-Almeida, B. P. Mosqueiro, & D. Bhugra (Eds.), *Spirituality and mental health across cultures—Evidence-based implications for clinical practice*. Oxford University Press. pp.11–25

References

Nigosian, S. A. (1993). *The Zoroastrian faith: Tradition and modern research*. McGill-Queen's University Press.

Obst, H. (2009). *Reinkarnation: Weltgeschichte einer Idee*. Verlag C.H. Beck. https://www.chbeck.de/obst-reinkarnation/product/26017

Ogden, D. (2001). *Greek and Roman necromancy*. Princeton University Press.

Otto, R. (1958). *The idea of the holy*. Oxford University Press.

Plato. (1913). *I. Eutyphro, Apology, Crito, Phaedo, Phaedrus. H. N. Fowler (Trans.)*. William Heinemann.

Rohde, E. (2010). *Psyche: Seelencult und Unsterblichkeitsglaube der Griechen. Volume 1*. Cambridge University Press. https://doi.org/10.1017/CBO9780511710681

Scholz, H. (1934). *Goethes Stellung zur Unsterblichkeitsfrage*. Mohr.

Shushan, G. (Ed.). (2009). *Conceptions of the afterlife in early civilizations: Universalism, constructivism, and near-death experience*. Continuum.

Sockness, B., & Gräb, W. (2010). *Schleiermacher, the study of religion and the future of theology*. De Gruyter.

Sommer, A. (2013). Crossing the boundaries of mind and body. Psychical research and the origins of modern psychology. In *Science and technology studies: Vol. PhD*. University College London.

Sommer, A. (2016). Are you afraid of the dark? Notes on the psychology of belief in histories of science and the occult. *European Journal of Psychotherapy and Counselling, 18*(2), 105–122. https://doi.org/10.1080/13642537.2016.1170062

Stark, R. (1999). Secularization, R.I.P. *Sociology of Religion, 60*(3), 249–273. https://doi.org/10.2307/3711936

VanderWeele, T., Balboni, T. A., & Koh, H. (2022). Invited commentary: Religious service attendance and implications for clinical care, community participation, and public health. *American Journal of Epidemiology, 191*(1), 31–35.

Wang, H. (1996). *A logical journey: From Gödel to philosophy*. MIT Press.

Weinreich, O. (1909). *Antike Heilungswunder*. Alfred Töpelmann.

3
Setting the Scene: Addressing the Main Arguments Against Survival Hypothesis

This chapter will present and discuss the most frequently claimed empirical and philosophical objections to survival. For those who argue that survival is unquestionably false and impossible, there is no need even to take a look at the empirical evidence claimed for survival, since they would *a priori* know that it is impossible. This would be reasonable, because, if survival is impossible, this alleged evidence must be generated by fraud, observational or inferential error, or some other flaw. Despite lacking significant proof or demonstration, these objections are often presented as "the scientific way of thinking" instead of metaphysical (and often ideological) claims, which they usually are.

Neuroscience "Proves" that the Brain Generates Mind

It is often taken for granted, both by the lay and academic public, that neuroscience has proved the physicalist view of mind, i.e., that mind is simply a product of the brain's chemical and electrical activity (Araujo, 2012; Gabriel, 2017; Holder, 2008; Moreira-Almeida et al., 2018). So, the argument goes, if the brain is a necessary condition for the mind, the mind cannot exist once the brain is dead. Usually, three kinds of empirical evidence showing that changes in the brain are associated with changes in the mind are presented as compelling proof thereof: (1) brain stimulation, drugs, brain lesions, or physiological dysfunctions affecting mind (e.g., generating hallucinations or memory loss); (2) correlations between brain and mind activities (e.g., feeling fear correlates to amygdala activation); and (3) parallels between evolution (or development) of brain and mind, i.e., increasingly higher mental function is exhibited by beings as they progress through life (from baby to adult) or evolution of the species (from lower animals to human beings) (Martin & Augustine, 2015).

While these three kinds of evidence are often taken for granted as conclusive proof that mind is just a product of brain activity, it is also constantly overlooked that these phenomena have been well-known to prominent psychologists and neuroscientists who have defended non-physicalist views of mind. Actually, the two key founders of modern scientific psychology, Wilhelm Wundt (Araujo, 2016) and William James (James, 2002), did not hold physicalist views of the mind. In

© The Author(s), under exclusive license to Springer Nature Switzerland AG 2022
A. Moreira-Almeida et al., *Science of Life After Death*, SpringerBriefs in Psychology, https://doi.org/10.1007/978-3-031-06056-4_3

3 Setting the Scene: Addressing the Main Arguments Against Survival Hypothesis

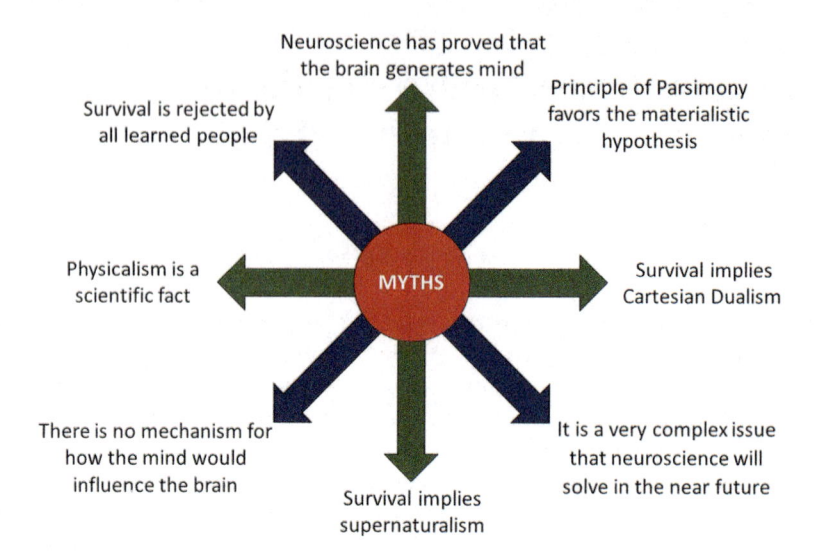

Fig. 3.1 Myths for assessing survival hypothesis

Diagram: Central circle labeled "MYTHS" with eight arrows pointing outward to:
- Neuroscience has proved that the brain generates mind
- Principle of Parsimony favors the materialistic hypothesis
- Survival implies Cartesian Dualism
- It is a very complex issue that neuroscience will solve in the near future
- Survival implies supernaturalism
- There is no mechanism for how the mind would influence the brain
- Physicalism is a scientific fact
- Survival is rejected by all learned people

addition, two of the major founders of neuroscience, Charles Sherrington and Wilder Penfield, performed long and in-depth investigations into the phenomena listed above, thought profoundly about their implications for the nature of mind and arrived at the opposite conclusion: mind is beyond the brain and could not just be explained by it. Sir Charles Sherrington, creator of the term "synapse" and the seminal author of our current understanding of neurons and their function (Burke, 2007), stated, at the end of his outstanding career that earned him a Nobel prize: "mind … escapes chemistry and physics" (Eccles et al., 1979, p. 131) and "that our being should consist of two fundamental elements offers I suppose no greater inherent improbability than that it should rest on one only" (Eccles et al., 1979, p. 182). To a similar conclusion reached the neuroscientist and neurosurgeon Wilder Penfield, who provided in vivo electrical stimulation in patients' brain during neurosurgeries and was seminal to the localization of brain's areas related to certain functions, given origin to the famous "Penfield's homunculus." In a book summarizing his final thoughts about the nature of mind, Penfield stated:

> I am forced to conclude that there is no valid evidence that either epileptic discharge or electrical stimulation can activate the mind. … For my own part, after years of striving to explain the mind on the basis of brain-action alone, I have come to the conclusion that it is simpler (and far easier to be logical) if one adopts the hypothesis that our being does consist of two fundamental elements. (Penfield et al., 1978, p. 78, 80)

A similar dualist conclusion was reached by the Nobel laureate neuroscientist John Eccles and the philosopher of science Karl Popper (Eccles et al., 1979; Penfield et al., 1978; Popper & Eccles, 1977). Naturally, presenting the conclusions of these most prominent neuroscientists does not prove that physicalism is false. However, it makes it clear that a physicalist conclusion based on the three phenomena listed above is not that straightforward. This is so because there is another main perspective to evaluate this evidence: brain as an instrument for mind manifestation, not as a mind producer. This hypothesis was well discussed over a century ago. As the philosopher, physician, and psychologist William James puts it: "Brain *can* be an organ for limiting and determining to a certain form a consciousness elsewhere produced" (James, 1898, p. 294).

In this perspective, the brain would be a "transmissive organ," with its imperfections and limitations, restricting the full manifestation of mind (James, 1898, p. 280). The brain would be similar to a TV set that does not generate the TV program but allows us to watch it. When we were children, we often thought that TV characters lived inside the TV set, only later we realized this is not the case. Let's apply this analogy to the three kinds of evidence often used to "prove" that brain generates mind: (1) electrical stimulation or damage to TV circuits may generate or affect the TV's sound and images, (2) there will be strict correlates between TV's circuit activation and images displayed, and (3) a better picture (even 4 K!) and sound correspond to the technology and quality of the TV set. However, neither of these three lines of evidence "prove" that the TV set is the ultimate source, the producer of the TV programs, nor that they could not be displayed through other devices (e.g., other TV sets, cell phones, computers, etc.). In fact, they only show that the TV set is important for exhibiting the TV program in a certain form (on that TV screen).

In summary, as stated by James (1898) over a century ago, the three pieces of evidence above can be equally explained by both the production (materialist) and the transmissive theories, so they do not permit testing if the brain is the mind's instrument or generator.[1] Empirical evidence to test and compare these two theories needs to be sought elsewhere, and this will be the subject of Chaps. 4 and 5.

Several misleading rhetorical strategies commonly used by some physicalists have been exposed by many scientists and philosophers. One is the recurrent use of misguided analogies and metaphors (e.g., brain producing mind is a purely physicochemical process like digestion, or is the same as a computer) to disguise the lack of convincing empirical evidence and an adequate understanding of the whole process by which the brain would generate mind (Araujo, 2012). In addition, what we actually know about the alleged neural mechanisms of the mind and their implications is often overstated (Moreira-Almeida et al., 2018). As expressed by the Nobel laureate neuroscientist Eccles:

---

[1] As stated by a contemporary of James, the philosopher C. S. Schiller: "if … a man loses consciousness as soon as his brain is injured, it is clearly as good an explanation to say the injury to the brain destroyed the mechanism by which the manifestation of consciousness was rendered possible, as to say that it destroyed the seat of consciousness" (Schiller, 1891, p. 295–296).

> There is a general tendency to overplay the scientific knowledge of the brain, which, regretfully, also is done by many scientists and scientific writers. For example, we are told that the brain 'sees' lines, angles (…) and that therefore we will soon be able to explain how a whole picture is 'seen' (…). But this statement is misleading. All that is known to happen in the brain is that neurons of the visual cortex are caused to fire trains of impulse in response to some specific visual input. (Popper & Eccles, 1977, p. 225)[2]

When physicalists are confronted by the lack of robust empirical evidence or a reasonable theory that would bridge the *explanatory gap* between brain activity and consciousness activities (e.g., *qualia*, reasoning, feelings, etc.) (Chalmers, 1995), they often claim it is a very complex issue that neuroscience will solve in the near future. This is a rhetorical strategy exposed as "promissory materialism" by Popper and Eccles (1977, p. 97). In a similar vein, the prominent philosopher Markus Gabriel (2017) recently exposed the "neurocentrism" that "preaches: the self is a brain" (p. 20) as an ideology based on "neuromythology" and "neuromania" (Gabriel, 2017).

Principle of Parsimony: We Should Explain Mind Solely on a Material Basis

This often-invoked argument claims we should accept the simpler hypothesis, i.e., we should explain mind and spiritual/anomalous experiences solely on material basis, instead of including an extra cause (e.g., a non-material aspect of the mind or of the universe) (Martin & Augustine, 2015). However, it is a misuse of the Principle of Parsimony (or Occam's razor) (Lycan, 2009), since this states we should prefer the simpler hypothesis only if both candidates have the same explanatory and heuristic power. Or, the classical formulation of Occam's razor: "Entities are not to be multiplied beyond necessity" (Baker, 2016). Evidence frequently requires us to assume multiple causes for a phenomenon, such as those of health and disease. It is no longer acceptable to believe they could be explained just by biological, psychological, or social factors in isolation, but we need a comprehensive and dynamic biopsychosocial model of health (Engel, 1977).

As discussed in the previous topic, the hypothesis of the brain as an instrument of the mind explains neuroscientific findings and, in addition, better explains voluntary acts and spiritual experiences. As stated by Schiller (1891, p. 295), "it will fit the facts alleged in favor of materialism equally well, besides enabling us to understand facts which materialism rejected as 'supernatural'."[3]

---

[2] A similar warning was issued by another neuroscientist/philosopher pairing: "the ascription of psychological – in particular, cognitive and cogitative – attributes to the brain is (…) a source of much (…) confusion (…) the great discoveries of neuroscience do not require this misconceived form of explanation" (Bennett & Hacker, 2003, p. 3–4).

[3] This was also clearly explained by William James: "The transmission-theory also puts itself in touch with a whole class of experiences that are with difficulty explained by the production-theory.

There Is No Mechanism for How the Mind Would Influence the Brain

It is desirable to have a theoretical mechanism to explain a given empirical finding, but it is not a *sine qua non* (Ohlsson & Kendler, 2020) in any science. For example, Sir Austin Bradford Hill, in his classic paper on the criteria for judging a causal connection in epidemiology, talks about the criteria of "plausibility":

> It will be helpful if the causation we suspect is biologically plausible. But this is a feature I am convinced we cannot demand. What is biologically plausible depends upon the biological knowledge of the day. … the association we observe may be one new to science or medicine and we must not dismiss it too light-heartedly as just too odd. As Sherlock Holmes advised Dr. Watson, 'when you have eliminated the impossible, whatever remains, *however improbable*, must be the truth.' (Hill, 1965, p. 298).

John Ioannidis, one of today's most prominent epidemiologists, recently reexamined the accumulated, empirical evidence for Hill's criteria in the 50 years since its publication and concluded that "In most cases, evocation of plausibility has been misleading" (Ioannidis, 2016, p. 1760–1761). To tell the truth, major scientific discoveries and their consequent theoretical advancements usually take place precisely when scientists observe phenomena they cannot explain and/or which do not conform to the established paradigms (Shapin, 1996). This is what the philosopher of science Thomas Kuhn called paradigm "anomalies" (Kuhn, 1970). Instead of being neglected, these anomalies should be investigated in depth; otherwise, it could generate a dogmatic denial of empirical evidence and hamper the progress of science (Chibeni & Moreira-Almeida, 2007). For example, until the early nineteenth century, it had been "known" for more than 2000 years (since Aristotle, and confirmed by Newton) "that no small bodies exist in space beyond the Moon" (Marvin, 2007, p. B10). Based on this, reports of meteorites were plainly rejected by most scientists (including some of the more prominent ones, such as Lavoisier) without any investigation because, since (as was believed) there are no rocks in the sky, they cannot fall from the sky, and this idea would contradict "common sense and the laws of physics" (Marvin, 2007, p. B10).

The preponderance of empirical evidence over theory was the core of the Scientific Revolution (Shapin, 1996), and it is even more necessary in a pre-paradigmatic phase in a Kuhnian sense.[4] This pre-paradigmatic phase is certainly where we are in terms of scientific exploration of mind. Often, a new paradigm or a new scientific field emerges proposing phenomenological theories, heavily based on

---

I refer to those obscure and exceptional phenomena reported at all times throughout human history […] religious conversions, providential leadings in answer to prayer, instantaneous healings, premonitions, apparitions at time of death, clairvoyant visions or impressions, and the whole range of mediumistic capacities […] All such experiences, quite paradoxical and meaningless on the production-theory, fall very naturally into place on the other theory. We need only suppose the continuity of our consciousness […]" (James, 1898, p. 298–299).

[4] A pre-paradigmatic phase is when a scientific field has not yet reached a paradigm, a consensus in the scientific community about theories and methods to explore a given group of phenomena.

observed phenomena, in spite of not being able to explain the mechanisms by which these phenomenological theories work. Disruptive scientific discoveries occurred counter to well-established theories and usually with no clear mechanism to explain the findings (Shapin, 1996). Darwin and Mendel proposed natural selection and genetics despite not being able to explain the intrinsic mechanisms (e.g., how the organism's traits were passed down to the next generation). This was possible only decades later, with the development of molecular biology leading to revolutionary discoveries such as DNA and genetic mutations (Chibeni & Moreira-Almeida, 2007).[5] Newton's law of gravitation, which requires action at a distance, contradicted head-on the prevailing mechanical principle, and he was accused of "reintroducing discredited occult powers" into science (Shapin, 1996, p. 42). It is useful to recall that physics still cannot explain the mechanisms by which gravity's action at a distance takes place.

To be fair, key scientists and philosophers have raised hypotheses to explain how a non-physical mind could influence brain function. For example, the prominent quantum physicist Henry Stapp proposed a "quantum interactive dualism" (Schwartz et al., 2005; Stapp, 2006). In a similar vein, a leading philosopher of religion proposed that the mind could influence wave function collapse in the physical body (Plantinga, 2011).[6] A collection of several other models developed to accommodate such mind-brain interactions is available in the book *Beyond Physicalism* (Kelly et al., 2015). The point here is not to argue which model is true but to show that several important authors have worked on theories that could account for mind-body interaction.

Regarding the oft-raised objection that the influence of a non-material mind over the brain would violate the principle of the conservation of energy, Plantinga reminds us that, even in classical physics, "this principle is stated for closed systems; but any physical system (a brain, e.g.) in which an immaterial substance caused a change would obviously not be a closed system" (Plantinga, 2011). Since the brain is not an isolated or closed system but receives an external force from the mind, the total amount of energy changes and does not therefore violate this law of classical physics (Averill & Keating, 1981).

In addition, as stated by the philosopher of science Robert Almeder (1992), discussing precisely the evidence for survival, we may know *that* something happens, even if we do not know *how* or *why* it happens. It is also good to remember that even ardent proponents of physicalist views of the mind acknowledge that they are not even close to a good theory to explain how the brain would generate mind (Araujo,

---

[5] Similar examples would be the development of the germ theory to explain infectious diseases that used to be regarded as caused by miasmas. Ignaz Semmelweis was able to explain death by puerperal sepsis (Noakes et al., 2008) and John Snow the cholera outbreak in London (Vandenbroucke et al., 1991) with rudimentary versions of the germ theory, despite not being able to show the bacteria that caused them, much less the mechanisms by which bacteria live, reproduce, and cause harm to human beings.

[6] Examples of other proposals are the "no-collapse" model proposed by the physicist Chris Clarke (2012) and the "quantum soul" hypothesis by the anesthesiologist Stuart Hameroff (Clarke, 2012).

2012; Blackmore, 2006). In summary, the alleged absence of a good explanatory theory (or reports of empirical phenomena that allegedly contradict established theories) is not a good reason to reject non-physicalist views of mind and its survival of bodily death. This is especially true in the pre-paradigmatic period we live in regarding the nature of mind and its relationship to the brain (Chibeni & Moreira-Almeida, 2007).

Science Has Proved Physicalism and Survival Implies Supernaturalism

It is often stated that, because science has proved physicalism, there must be a perennial and inevitable conflict between science and religion and all learned people are physicalists and deny survival. Only the superstitious would maintain these beliefs since they would not take into consideration other possible explanations: "The investigators [of psychic phenomena] … were believers, and they made no effort to rule out fraud" (Lester, 2005, p. 198).

This line of argument combines misguided philosophical reasoning, historical myths, strawman strategy, and *ad hominem* attacks. Firstly, physicalism is a metaphysical assumption (Stoljar, 2017). As a philosophical interpretation about the knowledge provided by scientific research, it is not and cannot be a scientific fact in itself. It is beyond the scope of science to tell of all that might exist in the universe, but metaphysical speculation can build consistent hypotheses about what is being this scope. Physicalism and spiritualism are examples of such general metaphysical hypothesis. It is also important to distinguish science (a method based on rational and empirical investigation of nature) from materialist scientism (an ideology that is disguisedly presented as scientific fact) (Haught, 2005). The sciences sometimes describe phenomena that suggest alternative or contradictory philosophical consequences. Some phenomena may suggest or allow a more materialist interpretation of nature, while other ones may suggest "purpose" (Coelho, 2022).

Secondly, in addition to physicalism being an additional metaphysical assumption, it is not a requirement for scientific inquiry nor is it a consequence of the scientific study of nature (Plantinga, 2011). In fact, by and large, most of the founders of modern science (e.g., Francis Bacon, Robert Boyle, Isaac Newton, Johannes Kepler, and René Descartes) were not physicalists but typically, in fact, religiously motivated to pursue science. They thought that science, as a study of creation, would help us to get to know the creator better and strengthen religious faith. In addition, it is currently a well-established historical fact that the idea of an inevitable and perennial conflict between science and religion is a myth (Alvarado, 2013; Harrison, 2015; Josephson, 2017; Numbers, 2009; Plantinga, 2011; Shapin, 1996; Sommer, 2013). Recent surveys have found a high prevalence of beliefs in survival and/or non-physicalist views of the human being among scientists, philosophers, psychiatrists, and people with a higher education (Bourget & Chalmers, 2014; Curcio &

Moreira-Almeida, 2019; Menegatti-Chequini et al., 2019; Moreira-Almeida & Araujo, 2015).

Thirdly, it is not historically accurate to describe survival researchers as naïve believers who have not considered the possibility of fraud or other alternative hypotheses. Actually, many of the leading scientific minds since the nineteenth century have been involved in survival research—William James, Alfred Russel Wallace, Cesare Lombroso, William Crookes, Oliver Lodge, Pierre Janet, Carl Jung, Theodore Flournoy, and Hans Eysenck (Moreira-Almeida, 2012)—as well as several Nobel laureates, Charles Richet, Pierre Curie and Marie Curie, J. J. Thomson, Henri Bergson, and Lord Rayleigh (Stevenson, 1977). Cardeña (2015) has compiled a list of more than 200 eminent scientists and intellectuals who took seriously the possibility of psi phenomena. Many of them did not deal directly with survival, but it shows that psi research is not just a matter for less intelligent people.

A landmark in the scientific investigation of survival was the founding of the Society for Psychical Research (SPR) in 1882, headed up by fellows from Trinity College Cambridge and composed of highly qualified scientists and intellectuals (Gauld, 1968; West, 2015).[7] It is also worth noting that many SPR researchers and other key figures in the field did not believe in survival (Gauld, 1968; Sommer, 2013). In addition, several SPR researchers started their studies as skeptics of survival and changed their minds based on the studies they performed, e.g., William Crookes, Cesare Lombroso, Oliver Lodge, Richard Hodgson, James Hyslop, and Frederic Myers (Hyslop, 1901; Lodge, 1929; Lombroso, 1909; Moreira-Almeida, 2006). Good researchers who accept the survival hypothesis typically first test and rule out the "usual suspects," i.e., the main conventional explanations "namely, fraud, misreporting, malobservation, or hidden memories (cryptomnesia)" (Almeder, 1992; Braude, 2003, p.10; Stevenson, 1977). As William James put it:

> In fact, were I asked to point to a scientific journal where hard-headedness and never-sleeping suspicion of sources of error might be seen in their full bloom, I think I should have to fall back on the *Proceedings of the Society for Psychical Research*. (Murphy & Ballou, 1960, p. 29)

Finally, the acceptance of a non-physical mind and its survival does not imply denying or neglecting all the scientific achievements humanity has achieved to date. In fact, this new knowledge about human beings would add to, and not replace, what we already know through science. As often commented by the psychiatrist Ian Stevenson, to explain human behavior, in addition to "Nature" (biology) and "Nurture" (environment), he would add a non-physical and surviving "Mind" (Kelly, 2013). So, the purpose is not to replace knowledge but to expand it, taking more factors into consideration for a more comprehensive view of human beings.

---

[7] The SPR's aim was: "to approach these various problems [psychical and spiritualistic phenomena] without prejudice or prepossession of any kind, and in the same spirit of exact and unimpassioned inquiry which has enabled Science to solve so many problems ... The founders of this Society fully recognize the exceptional difficulties which surround this branch of research; but they nevertheless hope that by patient and systematic effort some results of permanent value may be attained" (Sidgwick, 1882, p. 4).

Usually, these authors do not endorse supernaturalism but rather an expanded naturalism that regards consciousness as an irreducible component of the universe, in addition to matter, but also subject to natural laws (Kelly, 2007; Myers, 1903; Nagel, 2012; Schiller, 1891; Wallace, 1866).

Survival Implies Cartesian Dualism that Is Rejected by Learned People

Firstly, there are several kinds of substance dualism, and Cartesian dualism[8] is just one of them (Loose et al., 2018). Secondly, dualism is not the only non-physicalist theoretical option to explain mind. For example, idealism and panpsychism are two other common perspectives (Kelly et al., 2015). Finally, attacks on Cartesian dualism have been a paradigmatic example of strawman strategy, the misrepresentation of the opponent's ideas to make it easier to attack and defeat[9] (Duncan, 2000; Loose et al., 2018; Moreira-Almeida et al., 2018). Actually, Descartes proposed an interactionist dualism that acknowledges a complex and intricate, bidirectional influence between mind and body, allowing both mind to influence the body and body changes to influence the mind. It is worth reading the paper "Giving Dualism Its Due," written by the materialist philosopher William Lycan (2009) in which, in "an uncharacteristic exercise in intellectual honesty" (p. 551), he analyses the standard objections to dualism and concludes that:

> no convincing case has been made against substance dualism, and that standard objections to it can be credibly answered … going just by actual arguments as opposed to appeals to decency and what good guys believe, materialism is not significantly better supported than dualism. (Lycan, 2009, p. 551,562)

The claim that dualism is no longer accepted by any learned person is also patently false. Recent surveys among Brazilian psychiatrists (Menegatti-Chequini et al., 2019; Moreira-Almeida & Araujo, 2015), European university students and health professionals (Demertzi et al., 2009), and philosophers in Europe and North America (Bourget & Chalmers, 2014) show a high frequency of acceptance of dualism. In addition, substance dualism is still a major player in contemporary philosophy of mind as shown by recent works on the topic by prominent philosophers (Almeder, 2022; Plantinga, 2011; Swinburne, 2013) and a recent companion book on substance dualism with contributions from dozens of philosophers (Loose et al., 2018).

---

[8] Substance dualism in philosophy of mind proposes we are composed of two fundamental and radically different kinds of things: mind and body. Cartesian dualism is a very influential form of dualism proposed by René Descartes, claiming that mind (*res cogitans*) was a totally immaterial substance that has no extension in space (Robinson, 2020).

[9] For example, that Cartesian dualism would propose mind and brain as compartmentalized, non-interacting parts of human beings denying psychosomatic influences (Duncan, 2000; Loose et al., 2018; Moreira-Almeida et al., 2018).

Actually, these preliminary objections to survival are far weaker than often claimed by their supporters, and, regrettably, they have frequently led to a dogmatic materialist scientism. In summary, there is no sound argument or empirical evidence to compel us to an *a priori* rejection of survival as an explanatory hypothesis for the anomalous and spiritual experiences we will discuss in this book (Hyslop, 1901; Plantinga, 2011). In light of this, the survival hypothesis should be taken into consideration through a rigorous but open-minded and fair examination. Refusing to consider this possibility would be dogmatic and, thus, anti-scientific.

References

Almeder, R. F. (1992). *Death and personal survival: The evidence for life after death*. Rowman & Littlefield Publishers.
Almeder, R. F. (2022) *Materialism, Minds, and Cartesian Dualism*. Hamilton Books.
Alvarado, C. S. (2013). Psychic phenomena and the mind-body problem: Historical notes on a neglected conceptual tradition. *Archives of Clinical Psychiatry, 40*(4), 157–161. https://doi.org/10.1590/S0101-60832013000400006
Araujo, S. F. (2016). *Wundt and the philosophical foundations of psychology: A reappraisal*. Springer. https://doi.org/10.1007/978-3-319-26636-7
Araujo, S. F. (2012). Materialism's eternal return: Recurrent patterns of materialistic explanations of mental phenomena. In A. Moreira-Almeida & F. Santana Santos (Eds.), *Exploring Frontiers of the mind-brain relationship* (pp. 3–15). Springer. https://doi.org/10.1007/978-1-4614-0647-1_1
Averill, E., & Keating, B. F. (1981). Does Interactionism Violate a Law of Classical Physics? *Mind, 90*(357), 102–107.
Blackmore, S. (2006). *Conversations on consciousness*. Oxford University Press.
Baker, A. (2016). The Stanford Encyclopedia of philosophy. In E. N. Zalta (Ed.), *Simplicity*. https://plato.stanford.edu/archives/win2016/entries/simplicity/
Bennett, M. R., & Hacker, P. M. S. (2003). *Philosophical foundations of neuroscience*. Blackwell.
Bourget, D., & Chalmers, D. J. (2014). What do philosophers believe? *Philosophical Studies, 170*(3), 465–500. https://doi.org/10.1007/s11098-013-0259-7
Braude, S. E. (2003). *Immortal remains: The evidence for life after death*. Rowman & Littlefield.
Burke, R. E. (2007). Sir Charles Sherrington's the integrative action of the nervous system: A centenary appreciation. *Brain, 130*(Pt 4), 887–894. https://doi.org/10.1093/brain/awm022
Cardeña, E. (2015). *Eminent People Interested in Psi. In Psi Encyclopedia* (https://psi-encyclopedia.spr.ac.uk/articles/eminent-people-interested-psi). The Society for Psychical Research.
Chalmers, D. J. (1995). Facing up to the problem of consciousness. *Journal of Consciousness Studies, 2*, 200–219. https://doi.org/10.1093/acprof:oso/9780195311105.003.0001
Chibeni, S. S., & Moreira-Almeida, A. (2007). Remarks on the scientific exploration of "anomalous" psychiatric phenomena. *Archives of Clinical Psychiatry, 34*, 8–16. https://doi.org/10.1590/S0101-60832007000700003
Clarke, C. (2012). No-collapse physics and consciousness. In A. Moreira-Almeida & F. Santana Santos (Eds.), *Exploring Frontiers of the mind-brain relationship* (pp. 55–78). Springer. https://doi.org/10.1007/978-1-4614-0647-1_4
Coelho, H. S. (2022). The rationality of beauty: Aesthetics and the renaissance of teleology. *Zygon, 57*(1), 46–59.
Curcio, C. S. S., & Moreira-Almeida, A. (2019). Who does believe in life after death? Brazilian data from clinical and non-clinical samples. *Journal of Religion and Health, 58*(4), 1217–1234. https://doi.org/10.1007/s10943-018-0723-y

References

Demertzi, A., Liew, C., Ledoux, D., Bruno, M. A., Sharpe, M., Laureys, S., & Zeman, A. (2009). Dualism persists in the science of mind. *Annals of the New York Academy of Sciences, 1157*, 1–9. https://doi.org/10.1111/j.1749-6632.2008.04117.x

Duncan, G. (2000). Mind-body dualism and the biopsychosocial model of pain: What did Descartes really say? *The Journal of Medicine and Philosophy, 25*(4), 485–513. https://doi.org/10.1076/0360-5310(200008)25:4;1-A;FT485

Eccles, J. C., Gibson, W. C., & Sherrington, C. S. (1979). *Sherrington: His life and thought*. Springer International.

Engel, G. L. (1977). The need for a new medical model: A challenge for biomedicine. *Science, 196*(4286), 129–136. https://doi.org/10.1126/science.847460

Gabriel, M. (2017). *I am not a brain: Philosophy of mind for the 21st century*. Polity.

Gauld, A. (1968). *The founders of psychical research*. Schocken Books.

Harrison, P. (2015). *The territories of science and religion*. The University of Chicago Press.

Haught, J. F. (2005). Science and scientism: The importance of a distinction. *Zygon, 40*(2), 363–368. https://doi.org/10.1111/j.1467-9744.2005.00668.x

Hill, A. B. (1965). The environment and disease: Association or causation? *Proceedings of the Royal Society of Medicine, 58*(5), 295–300. https://doi.org/10.1177/003591576505800503

Holder, R. D. (2008). *Nothing but atoms and molecules? Probing the limits of science* (011052379) (3rd ed.). Faraday Institute for Science and Religion.

Hyslop, J. H. (1901). A further record of observations of certain trance phenomena. *Proceedings of the Society for Psychical Research, 16*, 1–649.

Ioannidis, J. P. A. (2016). Exposure-wide epidemiology: Revisiting Bradford Hill. *Statistics in Medicine, 35*(11), 1749–1762. https://doi.org/10.1002/sim.6825

James, W. (1898). Human immortality: Two supposed objections to the doctrine. In G. Murphy & R. O. Ballou (Eds.), *William James on psychical research* (pp. 279–308). Viking Press.

James, W. (2002). *Varieties of religious experience: A study in human nature*. Routledge. Publisher description. http://www.loc.gov/catdir/enhancements/fy0650/2002026869-d.html

Josephson, J. A., & nanda. (2017). *The myth of disenchantment: Magic, modernity, and the birth of the human sciences* (11272732). The University of Chicago Press. https://ebookcentral.proquest.com/lib/uchicago/detail.action?docID=4827749

Kelly, E. F. (2007). *Irreducible mind: Toward a psychology for the 21st century*. Rowman & Littlefield.

Kelly, E. F., Crabtree, A., & Marshall, P. (2015). *Beyond physicalism: Toward reconciliation of science and spirituality (18342882)*. Rowman & Littlefield.

Kelly, E. W. (2013). *Science, the self, and survival after death: Selected writings of Ian Stevenson* (E. W. Kelly, Ed.). Rowman & Littlefield Publishers. https://rowman.com/ISBN/9781442221147/Science-the-Self-and-Survival-after-Death-Selected-Writings-of-Ian-Stevenson

Kuhn, T. S. (1970). *The structure of scientific revolutions* (second edition, enlarged). University of Chicago Press.

Lester, D. (2005). *Is there life after death?: An examination of the empirical evidence*. McFarland.

Lodge, O. (1929). *Why I believe in personal immortality* (5th ed.). Cassell and company, ltd.

Lombroso, C. (1909). *After death—What? Spiritistic phenomena and their interpretation*.

Loose, J., Menuge, A. J. L., & Moreland, J. P. (2018). *The Blackwell companion to substance dualism* (021174972). Wiley Blackwell.

Lycan, W. G. (2009). Giving dualism its due. *Australasian Journal of Philosophy, 87*(4), 551–563. https://doi.org/10.1080/00048400802340642

Martin, M., & Augustine, K. (2015). *The myth of an afterlife: The case against life after death*. Rowman & Littlefield.

Marvin, U. B. (2007). Ernst Florens Friedrich Chladni (1756–1827) and the origins of modern meteorite research. *Meteoritics & Planetary Science, 42*(S9), B3–B68. https://doi.org/10.1111/j.1945-5100.2007.tb00606.x

Menegatti-Chequini, M. C., Maraldi, E. d. O., Peres, M. F. P., Leão, F. C., & Vallada, H. (2019). How psychiatrists think about religious and spiritual beliefs in clinical practice: Findings from a university hospital in São Paulo, Brazil. *Brazilian Journal of Psychiatry, 41*(1), 58–65. https://doi.org/10.1590/1516-4446-2017-2447

Moreira-Almeida, A. (2006). Book review—Is there life after death? An examination of the empirical evidence—David Lester. *Journal of Near-Death Studies, 24*(4), 245–254. https://doi.org/10.17514/JNDS-2006-24-4-p245-254

Moreira-Almeida, A. (2012). Research on mediumship and the mind–brain relationship. In *Exploring frontiers of the mind-brain relationship* (pp. 191–213). Springer.

Moreira-Almeida, A., & Araujo, S. F. (2015). Does the brain produce the mind? A survey of psychiatrists' opinions. *Archives of Clinical Psychiatry, 42*(3), 74–75. https://doi.org/10.1590/0101-60830000000051

Moreira-Almeida, A., Araujo, S. F., & Cloninger, C. R. (2018). The presentation of the mind-brain problem in leading psychiatry journals. *Braz J Psychiatry, 40*(3), 335–342. https://doi.org/10.1590/1516-4446-2017-2342

Murphy, G., & Ballou, R. O. (1960). *William James on psychical research*. Viking Press.

Myers, F. W. H. (1903). *Human personality and its survival of bodily death* (Vol. 1–2). Green and co.

Nagel, T. (2012). *Mind and cosmos: Why the materialist neo-Darwinian conception of nature is almost certainly false*. Oxford University Press.

Noakes, T. D., Borresen, J., Hew-Butler, T., Lambert, M. I., & Jordaan, E. (2008). Semmelweis and the aetiology of puerperal sepsis 160 years on: An historical review. *Epidemiology and Infection, 136*(1), 1–9. https://doi.org/10.1017/S0950268807008746

Numbers, R. L. (2009). *Galileo goes to jail: And other myths about science and religion*. Harvard University Press.

Ohlsson, H., & Kendler, K. S. (2020). Applying causal inference methods in psychiatric epidemiology: A review. *JAMA Psychiatry, 77*(6), 637–644. https://doi.org/10.1001/jamapsychiatry.2019.3758

Penfield, W., Hendel, C. W., Feindel, W., & Symonds, C. (1978). *The mystery of the mind: A critical study of consciousness and the human brain*. Princeton University Press.

Plantinga, A. (2011). *Where the conflict really lies: Science, religion, and naturalism*. Oxford University Press.

Popper, K. R., & Eccles, J. C. (1977). *The self and its brain*. Springer International.

Robinson, H. (2020). *"Dualism", the Stanford Encyclopedia of philosophy*. Edward N. Zalta.

Schiller, F. C. S. (1891). *Riddles of the sphinx: A study in the philosophy of evolution*. S. Sonnenschein.

Schwartz, J. M., Stapp, H. P., & Beauregard, M. (2005). Quantum physics in neuroscience and psychology: A neurophysical model of mind-brain interaction. *Philosophical Transactions of the Royal Society of London. Series B, Biological Sciences, 360*(1458), 1309–1327. https://doi.org/10.1098/rstb.2004.1598

Shapin, S. (1996). *The scientific revolution*. University of Chicago Press.

Sidgwick, H. (1882). Objects of the society. In *Proceedings of the Society for Psychical Research* (Vol. 1).

Sommer, A. (2013). Crossing the boundaries of mind and body. Psychical research and the origins of modern psychology. In *Science and technology studies: Vol. PhD*. University College London.

Stapp, H. P. (2006). Quantum interactive dualism: An alternative to materialism. *Zygon, 41*(3), 599–616. https://doi.org/10.1111/j.1467-9744.2005.00762.x

Stevenson, I. (1977). Research into the evidence of man's survival after death: A historical and critical survey with a summary of recent developments. *The Journal of Nervous and Mental Disease, 165*(3), 152–170. https://doi.org/10.1097/00005053-197709000-00002

Stoljar, D. (2017). Physicalism. In E. N. Zalta (Ed.), *The Stanford Encyclopedia of Philosophy* (Winter 2017 Edition). https://plato.stanford.edu/archives/win2017/entries/physicalism/

Swinburne, R. (2013). *Mind, brain, and free will*. Oxford University Press.

References

Vandenbroucke, J. P., Eelkman Rooda, H. M., & Beukers, H. (1991). Who made John snow a hero? *American Journal of Epidemiology, 133*(10), 967–973. https://doi.org/10.1093/oxfordjournals.aje.a115816

Wallace, A. R. (1866). *The scientific aspect of the supernatural: Indicating the desirableness of an experimental enquiry by men of science into the alleged powers of clairvoyants and mediums.* http://books.google.com/books?vid=BL:A0018283198Google_Books.

West, D. (2015). Society for Psychical Research | psi Encyclopedia. https://psi-encyclopedia.spr.ac.uk/articles/society-psychical-research.

4
What Would Constitute Evidence for Personal Survival After Death?

If we are searching for empirical evidence for a given hypothesis, we have to consider that the truth value of every proposition about observations is ultimately fallible, so that we will never be in a position to justify unrestricted certainty or unrestricted disbelief in any specific fact (Almeder, 1992). The best method to sustain the critical and educated balance between belief and disbelief, between dogmatic conviction and dogmatic skepticism, is *science*. But no science is based on a single, extremely persuasive piece of evidence.

But how exactly is a science born? What kinds of struggle does it have to endure to move from initial rejection to widespread acceptance? History and philosophy of science have been addressing this quandary for some time, and what we have learned from these reflections is intellectual humility, since more than once we have been mistaken about claims erstwhile deemed a certainty. Empirically based claims are never conclusive.[1] Some given experiences may suggest that the mind does not survive bodily death, while other experiences may suggest that it does.

At least since Edward Tylor in the nineteenth century, many scholars of religion have proposed that religious and spiritual beliefs are based on spiritual experiences (SE) since, at first sight, they suggest a non-material aspect of the human being and its survival (e.g., out-of-body experiences, seeing and hearing spirits of the dead, etc.) (Bozzano, 1997; Hardy, 1979; James, 2002; Lang, 1898; Rennie, 2008; Tylor, 1871). Actually, SE have been used since antiquity as evidence for a spiritual aspect of man for many of the most prominent scholars (e.g., Pythagoras, Socrates, and Plato – (Firth, 1904; Plato, 1913)), passing through the Middle Ages (most authors, but Thomas Aquinas in particular is worth mentioning (Aquinatis, 1880, [STh Ia, q75])) and the Modern Age (e.g., Robert Boyle, Rene Descartes, and Francis Bacon) (Josephson, 2017; Kripal, 2010; Sommer, 2013).

---

[1] As the wide philosophical tradition from Hume, Kant, and Coleridge to Popper shows. According to Coleridge's *Opus Magnum*, one is never completely serious about asserting that all sheep are white, for it is conceivable that some new experience disproves this assertion.

In the last two centuries, based on materialist metaphysics, any non-physicalist factor contributing to the explanation of SE has often, *a priori*, been rejected. Based on the metaphysical commitment to physicalism, all SE must be explained away as mental disorder, psychological distortions, social constructions, or physiological effects (Holder, 2008; James, 2002; Moreira-Almeida et al., 2005; Myers, 1903). This perspective gave birth to the principle of the Methodological Exclusion of the Transcendent, proposed by the Swiss psychologist Theodore Flournoy (1902), which proposes a "biological interpretation of religious phenomena," its "psycho-physiological nature" (Flournoy, 1902, p. 57).

However, many authors reject Methodological Exclusion of the Transcendent (Coelho, 2021) and defend the inclusion of the transcendent, at least as a hypothesis to be taken into consideration for an understanding and explanation of SE (Beauregard et al., 2014; Cardeña, 2014; Hardy, 1979; Hood et al., 2018; Kelly, 2007, 2013; Kripal, 2010; Lang, 1898; Moreira-Almeida & Santos, 2012; Myers, 1903; Sech Junior et al., 2013; Stark, 2017; Walach, 2014). What they defend is not a denial of the biopsychosocial explanatory factors of SE. These factors can explain much (perhaps most), but not all SE, for example, when an NDE or a visionary experience conveys veridical information unknown to the observer (Stevenson, 1983). Why must these experiences necessarily be explained only by physical factors and why should we be forbidden from even entertaining the hypothesis that consciousness was actually beyond the body and was able to perceive such things? The crucial point is not that we should assume a non-physical explanation, but we should not dogmatically exclude it a priori based on previous metaphysical or ideological commitment.[2]

But what would be evidence for survival? Before discussing evidence for personal survival, it is essential to discuss personal identity and how we can know the existence of other minds (in addition to our own minds). Though we do not usually think about this, we can have direct access and evidence only about our own minds. We only directly experience our minds: feelings, thoughts, desires, sense of being, etc. As stated by René Descartes: *cogito, ergo sum* (I think, therefore I am).[3] However, how can I know that there are other minds, other thinking beings in the world given that I cannot experience their minds directly? What if other people are just mindless flesh machines, zombies? We *do* know of the existence of other minds because we perceive other bodies behaving in ways indicating they have minds like ourselves. They behave and express themselves as if they also think, feel, have desires and sense of being, etc. They also engage in meaningful communication

---

[2] The limiting nature of an a priori commitment to physicalism was well expressed by the scholar Chesterton: "Spiritual doctrines do not actually limit the mind as do materialistic denials. Even if I believe in immortality I need not think about it. But if I disbelieve in immortality I must not think about it. In the first case the road is open and I can go as far as I like; in the second the road is shut" (Chesterton, 1909, p. 40).

[3] In an even more powerful logical form, given by Fichte, *quodcunque cogitate, est*; whatever thinks testifies its own existence (Fichte, 1965, p. 262). Things that do not think, however, have to be proven, described, and justified by someone who thinks.

with us and are capable of shared attention, proving that we are interacting with "something" that is interpersonally aware of "seeing and being seen" (Hegel, 1970; Coelho, 2016).[4] But because we do not have direct access to their minds, we must base our verdict on indirect evidence, based on an observation of their behavior (Berkeley, 1975; Chibeni, 2020; Moreira-Almeida, 2012).

After acknowledging there are other minds out there, how can we decide about their identities? We can judge "the identity of a *mind* by a sufficiently rich set of its qualities: ideas, knowledge, desires, ways of reasoning and feeling, etc." (Chibeni, 2020, p. 4). A specific pattern of these qualities, a *continuity of character and memory*, is what characterizes a personality, an individual mind, or soul (Quinton, 2008).[5] For example, my friend Bob has a specific pattern of memories, feelings, values, ways to react to life events, humor, etc. that characterize him. Any time I meet him (i.e., I see his body), I expect such a cluster of personal features. However, what ultimately matters is not the body itself but the mental and psychological properties (Perry, 2008). According to the philosopher Lord Quinton, the body functions as "convenient recognition devices" for a given soul (personal identity).[6]

A thought experiment will help to illustrate this. Let's suppose that my friend Bob travels abroad and suffers a terrible accident, having his body entirely burned, but he survives, completely disfigured. If DNA testing is not available, how would I know that he is my dear friend Bob when I see him some months after the accident? I can no longer see features in his body that I used to recognize in my friend Bob. So, I'll have to rely entirely on the pattern of mental qualities that characterize him, i.e., I'll look for a psychological continuity of Bob's personality.

What if my friend Bob dies in the accident? What would constitute evidence for Bob's survival (his mind, personality, soul) after bodily death? I would have to look for the same continuity of character and memory that I had previously sought when he survived, but I could no longer use his body as a "recognition device" for his personality. So, empirical evidence of someone's personal survival after death would be evidence for persistence of the "thinking being," of the "I," after death or when the brain is not working. In order to recognize an alleged surviving personality after death, we would expect specific personal characteristics that would manifest themselves through the means they had at their disposal (e.g., a medium's body,

---

[4] Both idealistic research on intersubjectivity and contemporary analytic philosophical research on second-person perspective give fruitful insight on how our minds are in fact in an interdependent relationship with other minds—that is, *interpersonal* relationship. Research on psychosocial cognition has led to very similar results (Frith & Frith, 2007).

[5] As claimed by the philosopher Stephen Braude, "we decide who someone is on the basis of what they say and how they behave – more specifically, their memory claims and continuity of character" (Braude, 2003, p. 3–4).

[6] "What is unique about individual people that is important enough for us to call them by individual proper names? In our general relations with other human beings their bodies are for the most part intrinsically unimportant. We use them as convenient recognition devices enabling us to locate without difficulty the persistent character and memory complexes in which we are interested, which we love or like. (…)" (Quinton, 2008, p. 64).

or a new body in cases of reincarnation) (Almeder, 1992; Braude, 2003; Stevenson, 1977). For example:

- Memory
  - Being able to remember facts, ideally in large number, accurate, and covering several topics
  - Identifying people the claimed personality was acquainted with when alive
- Skills of the alleged personality
  - Speaking or writing in a foreign language
  - Artistic: poetry, prose, painting, playing some musical instrument, etc.
  - Handwriting
- Personality traits: temperament, character, and personal style (Moreira-Almeida, 2012, p. 196–197)

The best evidence for a surviving mind would not only be the persistence of memory, "a memory bank is not a person" (Gauld, 1982, p. 75). What is important is the persistence of the character, especially evidence that the "I" is still active: new verifiable memories and evidence of will, personal skills, affection, peculiar goals, humor, mannerisms, etc., regarding new events and situations.[7]

The evidence discussed above, if real, would invalidate physicalist views of consciousness and human nature. For a given hypothesis to be scientific and not dogmatic, it must be possible to imagine possible empirical evidence that would falsify[8] it (Moreira-Almeida, 2006). Making these physicalist views falsifiable in the sense proposed by the philosopher of science Karl Popper (1963): if consciousness is generated and depends on the brain functioning for its existence, there should not be any evidence of mind activity when the brain is dead or not functional (Popper, 1963). If this sort of evidence is consistently found, especially by different investigators using different methods and investigating different phenomena, they would falsify physicalist views of mind. The Chap. 5 will present and discuss the main empirical evidence for mind activity when the brain is not functional or dead.

---

[7] As stated by Alan Gauld: "evidence that certain deceased persons have after their deaths continued to attempt to pursue goals and purposes that were characteristic of them in life, or have begun to pursue goals that might be thought a natural development of these" (Gauld, 1982, p. 75).

[8] Karl Popper, one of the main philosophers of science of the twentieth century, proposed the criterion of falsifiability for demarcating science from non-science. According to this very influential criterion, a hypothesis is scientific only when it is possible to imagine some sort of empirical observation that potentially would falsify (prove false) that hypothesis (Popper, 1963).

References

Almeder, R. F. (1992). *Death and personal survival: The evidence for life after death*. Rowman & Littlefield Publishers.

Aquinatis, T. (1880). *Summa Theologica*. Bloud et Barral.

Beauregard, M., Schwartz, G. E., Miller, L., Dossey, L., Moreira-Almeida, A., Schlitz, M., Sheldrake, R., & Tart, C. (2014). Manifesto for a post-materialist science. *Explore, 10*(5), 272–274. https://doi.org/10.1016/j.explore.2014.06.008

Berkeley, G. (1975). *Philosophical works*. Everyman.

Bozzano, E. (1997). *Povos primitivos e manifestações supranormais*. Editora Jornalistica FE.

Braude, S. E. (2003). *Immortal remains: The evidence for life after death*. Rowman & Littlefield.

Cardeña, E. (2014). A call for an open, informed study of all aspects of consciousness. *Frontiers in Human Neuroscience, 8*, 17. https://doi.org/10.3389/fnhum.2014.00017

Chesterton, G. K. (1909). *Orthodoxy* (4054793). John Lane.

Chibeni, S. (2020). Spiritism: An experimental approach to the issue of personal post-mortem survival. *Jornal de Estudos Espíritas*. https://doi.org/10.22568/jee.v8.artn.010203

Coelho, H. S. (2016). A Constituição Histórica da Subjetividade nas Revisões da Teoria Hegeliana. *Síntese, 43*(136), 305–318.

Coelho, H. S. (2021). O que é e qual é o lugar do transcendente na filosofia? Uma nota a partir de G. W. F. Hegel. *Ética e Filosofia. Política, 23*(1), 155–162.

Fichte, J. G. (1965). *Gesamtausgabe der Bayerischen Akademie der Wissenschaften. 1,2: Werke 1793–1795*. Frommann-holzboog Verlag.

Firth, F. M. (1904). *The Golden verses of Pythagoras and other Pythagorean fragments*. Theosophical Publishing House.

Frith, C. D., & Frith, U. (2007). Social cognition in humans. *Current Biology, 17*(16), 724–732.

Flournoy, T. (1902). Les principes de la psychlogie religieuse. *Archives de Psychologie, 5*, 33–57.

Gauld, A. (1982). *Mediumship and survival: A century of investigations*. Heinemann.

Hardy, A. (1979). *The spiritual nature of man: A study of contemporary religious experience*. Clarendon Press.

Hegel, G. W. F. (1970). *Werke in 20 Bänden. Ed. Moldenhauer, Eva; Michel, Karl M.* Suhrkamp.

Holder, R. D. (2008). *Nothing but atoms and molecules? : Probing the limits of science* (3rd ed.). Faraday Institute for Science and Religion.

Hood, R. W., Hill, P. C., & Spilka, B. (2018). *The psychology of religion: An empirical approach* (5th ed.). Guilford Press.

James, W. (2002). *Varieties of religious experience: A study in human nature*. Routledge. Publisher description.

Josephson-Storm, J. A. (2017). *The myth of disenchantment: Magic, modernity, and the birth of the human sciences*. The University of Chicago Press.

Kelly, E. F. (2007). *Irreducible mind: Toward a psychology for the 21st century*. Rowman & Littlefield.

Kelly, E. W. (2013). In E. W. Kelly (Ed.), *Science, the self, and survival after death: Selected writings of Ian Stevenson*. Rowman & Littlefield Publishers.

Kripal, J. J. (2010). *Authors of the impossible: The paranormal and the sacred*. The University of Chicago Press.

Lang, A. (1898). *The making of religion*. Longmans, Green and Co.

Moreira-Almeida, A. (2006). Book review—Is there life after death? An examination of the empirical evidence—David Lester. *Journal of Near-Death Studies, 24*(4), 245–254. https://doi.org/10.17514/JNDS-2006-24-4-p245-254

Moreira-Almeida, A. (2012). Research on mediumship and the mind–brain relationship. In *Exploring frontiers of the mind-brain relationship* (pp. 191–213). Springer.

Moreira-Almeida, A., & Santos, F. S. (2012). *Exploring frontiers of the mind-brain relationship*. Springer. https://doi.org/10.1007/978-1-4614-0647-1

Moreira-Almeida, A., Silva de Almeida, A. A., & Neto, F. L. (2005). History of "Spiritist madness" in Brazil. *History of Psychiatry, 16*(61 Pt 1), 5–25. https://doi.org/10.1177/0957154X05044602

Myers, F. W. H. (1903). *Human personality and its survival of bodily death* (Vol. 1–2). Longmans, Green and co.

Perry, J. (Ed.). (2008). *Personal identity* (2nd ed.). University of California Press.

Plato. (1913). *I. Eutyphro, Apology, Crito, Phaedo, Phaedrus. H. N. Fowler (Trans.)*. William Heinemann.

Popper, K. R. (1963). *Conjectures and refutations: The growth of scientific knowledge*. Routledge & Kegan Paul.

Quinton, A. (2008). The soul. In J. Perry (Ed.), *Personal identity* (pp. 53–72). University of California Press.

Rennie, B. (2008). Mircea Eliade: 'Secular mysticism' and the history of religions. *Religion, 38*(4), 328–337. https://doi.org/10.1016/j.religion.2008.05.016

Sech Junior, A., Araujo, S. d. F., & Moreira-Almeida, A. (2013). William James and psychical research: Towards a radical science of mind. *History of Psychiatry, 24*(1), 62–78. https://doi.org/10.1177/0957154X12450138

Sommer, A. (2013). Crossing the boundaries of mind and body. Psychical research and the origins of modern psychology. In *Science and technology studies: Vol. PhD*. University College London.

Stark, R. (2017). *Why god?: Explaining religious phenomena*. Templeton Press.

Stevenson, I. (1977). Research into the evidence of man's survival after death: A historical and critical survey with a summary of recent developments. *The Journal of Nervous and Mental Disease, 165*(3), 152–170. https://doi.org/10.1097/00005053-197709000-00002

Stevenson, I. (1983). Do we need a new word to supplement "hallucination"? *The American Journal of Psychiatry, 140*(12), 1609–1611. https://doi.org/10.1176/ajp.140.12.1609

Tylor, E. B. (1871). *Primitive culture: Researches into the development of mythology, philosophy, religion, art, and custom*. J. Murray.

Walach, H. (2014). *Secular spirituality: The next step towards enlightenment*. Springer.

5
The Best Available Evidence for Life After Death

Mediumship

Mediums are individuals who claim to have the ability to perceive, communicate with, or be under the control of deceased persons (Moreira-Almeida, 2012). This experience has been universally reported throughout the ages and has been amply investigated. In the face of both religious and materialist dogmatism, leading intellectuals advocated the need for a serious investigation of such seemingly extraordinary phenomena.

Communication with the deceased and their apparitions have been very common and universal occurrences across cultures since the first records of civilization, though in accordance with the specific symbolism and metaphors of different religious and artistic traditions (Bozzano, 1997; Du Prel, 1889; Myers, 1903). Ancient Egyptians had the common practice of exchanging messages between the living and the dead (Helck & Otto, 1975, Vol. I, p. 297). Origen, one of the church fathers, argued that Christ could not resurrect in the same physical body but had to make (or receive) a celestial body for himself (Bigg, 1886, p. 197–200). This "thesis on the apparition of the dead" would make Christ's resurrection the greatest ghost story of all times while taking it seriously. It is not in dispute that almost all saints, spiritual masters, seers, and religious traditions involve at least occasional contact between the spiritual and the physical realm and at least some perceptions of these spiritual realities. According to these metaphors, the souls may present themselves as spirits, phantoms, angels, gods, saints, bodhisattvas, djinns, or even undefined entities but always carrying meaning or having a message to deliver.[1]

---

[1] To mention just a few remarkable examples, Dante Alighieri reveals personal encounters with some sort of spirit or angel (Alighieri, 1993, p. 23–24), Giordano Bruno considers them commonsensical (Bruno et al., 1998, p. 111–116), and J. G. Hamann even argued that the Holy Spirit of the Christians' and Socrates' *daemon* had the same origin (Hamann, 1959) so that all contacts with spiritual reality were possible sources of revelation.

Although anecdotal evidence that convinced many observers has always existed, systematic, scientific studies only began in the second half of the nineteenth century (Gauld, 1982). Then, the idea of possible contact with departed souls became an object of empirical research, under different labels such as Spiritism, Modern Spiritualism, Animal Magnetism, and Psychical Research, among others.

A major figure in these initial efforts was the French scholar Hippolyte Léon Denizard Rivail, better known as Allan Kardec, who proposed an investigation into mediumistic phenomena through experimental methods and developed a philosophy called Spiritism (Chibeni, 2020; Moreira-Almeida, 2008). His purpose was to naturalize these experiences: "facts falsely qualified as supernatural are subject to laws, it makes them enter the order of the phenomena of nature" (Kardec, 1890).

A few decades later, several scientific societies were established to investigate such phenomena. The Society for Psychical Research (SPR) was founded in 1882 (Weaver, n.d.). The American Society for Psychical Research (ASPR) was founded under the leadership of William James, in 1884 (Gauld, 1982, Ch. 1). In France, the *Institut Métapsychique International* (IMI) was founded in 1919 by scholars including Charles Richet (Evrard, n.d.).

Mrs. Leonora Piper

Probably the best and most investigated medium at that time was Mrs. Leonora Piper, from Boston (Gauld, 1982, Ch. 3). The Harvard professor and founder of American scientific psychology, William James, conducted the first studies. He initially took 25 different sitters to her, all of whom used pseudonyms, and she presented correct information of private affairs related to each of the sitters. After more than 10 years' investigating Mrs. Piper, James was convinced that conventional explanations could not account for her trances. Mrs. Piper was James' "white crow":

> (…) If you wish to upset the law that all crows are black, you must not seek to show that no crows are; it is enough if you prove one single crow to be white. My own white crow is Mrs. Piper. In the trances of this medium, I cannot resist the conviction that knowledge appears which she has never gained by the ordinary waking use of her eyes and ears and wits. (…) (James, 1886, p. 131)

Mrs. Piper was continuously and rigorously investigated for almost 25 years by many scientists, including Sir Oliver Lodge, James Hyslop, and the skeptical "fraud hunter" Richard Hodgson (Gauld, 1982, Ch. 3). For example, she and her private contacts were followed by detectives to ensure she could not fraudulently obtain information. Also, she was taken from the USA to England, where she didn't know anyone, to be studied by SPR members. Even so, Mrs. Piper produced accurate information. Her ability was so impressive that she sometimes simultaneously wrote different messages on each hand, allegedly from different discarnate personalities, while dictating a third message (Gauld, 1982, Ch. 3; Hodgson, 1897). Most scientists who investigated her in depth became convinced she consistently obtained anomalous information, and many were persuaded that these came from discarnate personalities. Also, there was never any reasonable evidence for fraud despite being thoroughly investigated.

The material obtained through investigations in séances with Mrs. Piper is vast, and we present here just some illustrative examples. First, it is essential to note that the séances were conducted in daylight, with no special arrangement in the room. The researcher introduced the sitters anonymously (usually unknown to her), and she produced the mediumistic writing in full view of those present (Hyslop, 1901, Chps. 1 and 6). One illustrative example was the experiment conducted by Hodgson to test the identity of one of Mrs. Piper's "communicators"—George Pelham (GP). Hodgson invited GP's family members and friends and arranged several sessions with Mrs. Piper. In a state of trance, she easily recognized 29 out of 30 individuals invited among 120 other sitters unbeknown to GP, reporting pertinent knowledge about their private lives and engaging in conversation with them as GP would have done. The person she did not recognize was a lady GP had met when the lady was a little girl. GP was an acquaintance of her mother. After a brief comment about the sitter's mother, Mrs. Piper (in trance) said the words below, reporting specific knowledge about the lady's mother:

> Of course, oh, very well. For pity sake, are you her little daughter? By Jove, how you have grown ... I thought so much of your mother, a charming woman. Our tastes were similar. Do you know Marte at all? Your mother knows. Ask her if she remembers the book I gave her to read. And ask her if she still remembers me and the long talks we used to have at the home evenings. I wish I could have known you better, it would have been so nice to have recalled the past. (Hodgson, 1897, p. 324-325)

The first sitting relating to GP took place 5 weeks after his death and was conducted by Hodgson and assisted by an intimate friend of GP (introduced as Mr. Hart, a pseudonym). In a trance, Mrs. Piper gave GP's real name in full, and also those of "several of his most intimate friends, including the name of the sitter" (Hodgson, 1897, p. 297) and incidents unknown to the sitters. For example, Mr. and Mrs. James and May Howard were mentioned (GP's friend and their names were not mentioned by the sitter) as well as a message to their daughter Katharine: "Tell her, she'll know. I will solve the problems, Katharine" (Hodgson, 1897, p. 297). These words had no meaning to Mr. Hart. However, when told to Mrs. Howard, she was impressed because "when he [GP] had last stayed with them, he had talked frequently with Katharine (a fifteen-year-old girl) upon such subjects as Time, Space, God, Eternity, and pointed out to her how unsatisfactory the commonly accepted solutions were. He added that some time he would solve the problems, and let her know, using almost the very words of the communication made at the sitting" (Hodgson, 1897, p. 297-298).

Another intriguing kind of phenomenon produced by Mrs. Piper was the so-called cross-correspondence, in which different mediums simultaneously produce writings that independently have no meaning, but when combined, they fit like pieces of a puzzle. One example was "the Abt Vogler case," where Myers, a recently deceased scholar and SPR founder, supposedly sent complementary fragments of a message to three different mediums. First, in a séance with Mrs. Piper, held in England on January 16, 1907, SPR researcher Piddington asked the alleged Myers to send a similar message to the mediums Mrs. Verrall and Mrs. Holland, with a simple sign. He requested: "if you wrote, for instance, 'sunshine' through Mrs. Verrall and then afterwards through Mrs. Holland, you might put, say, a triangle within a circle, or some simple sign like that, to show that there is another message to be looked for, corresponding with the message thus marked" (Piddington, 1908, Section III). On January 29, a circle with a triangle inside appeared in Mrs. Verrall's writing.

This cross-correspondence case proceeded in the scripts produced by Mrs. Verrall, her daughter (Miss Verrall), and Mrs. Piper (Piddington, 1908, Section III). In a sitting held on February 11, the alleged Myers advised Piddington through Mrs. Piper's reading "… look out [on the Mrs. Verrall scripts] for Hope Star and Browning." Piddington found in Mrs. Verrall's scripts references to the poem Abt Vogler by Browning which contains the words "Hope, Star and Browning." The next day, the words "Evelyn Hope" were given by the alleged Myers to Mrs. Verrall—a title of one of Browning's poems. Moreover, the writing that mentioned hope, star, and Browning ended with a drawing of a circle with a triangle inside. Another interesting fact was that, in the poem, the original word "passion" had been replaced by "hope," and Mrs. Verrall wrote to Piddington "I knew perfectly when I read the script that it should have been 'passion' which left the ground for the sky - and I was annoyed at the blunder" (Piddington, 1908, Section VI, p. 63). So, even though she knew the poem, she wrote it "wrongly," but it matched Mrs. Piper's script. Finally, on February 17, Miss Verrall wrote a fragment including the word "star" that alluded to Browning's Pied Piper of Hamelin (Piddington, 1908, Section VI).

One may argue that Mrs. Piper's trances did also produced incorrect information and the veridical information was given at random. Indeed, failures occurred and, as is usual with any human skill, the quality fluctuated sharply (Gauld, 1982; Hodgson, 1897; Hodgson, 1892; Hyslop, 1905; Lodge, 1909). Nonetheless, up to the present day, the records produced by her investigation have never been exceeded in number and detail (Gauld, 1982, Ch. 3). The material produced was so impressive that after years of in-depth investigations into Mrs. Piper, Hodgson, initially a skeptic, concluded:

> After having endeavoured as best I could to follow the writing of thousands of pages, with scores of different writers, after having put many inquiries to the communicators themselves ... I cannot profess to have any doubt but that the chief 'communicators'... are veritably the personalities that they claim to be, that they have survived the change we call death. (Hodgson, 1897, p. 405-6)

Chico Xavier

Another very prolific medium and the subject of recent studies was the Brazilian Chico Xavier (1910–2002). He grew up in a very poor Brazilian family, only received an elementary education, and worked as a public servant until retirement.

Despite this, he produced more than 450 books through psychography (writing under the alleged influence of a deceased person), covering a wide range of precise information and different text styles such as letters, poems, novels, and scientific texts. Also, he has never accepted any kind of payment or gifts for his work, donating all copyrights to charity (his books sold more than 60 million copies, translated into 16 languages) (Playfair, 2010; Souto Maior, 2003). For decades he conducted séances twice a week, attended by dozens or hundreds of people, where he used to automatically write around six letters per séance attributed to sitters' deceased relatives (Rocha et al., 2014; Severino, 1992, Ch. 3).

Some of these letters were recently assessed scientifically by our research group. One study evaluated a letter attributed to JS, a first-year medical student who died in a train accident. The following year, JS' parents traveled 300 miles to attend Chico's public séances. The only information they gave to the medium was the names of JS and his mother. On their third visit, they received a letter containing 29 verifiable pieces of information that were rated as a "clear and precise fit," comprising highly specific data such as names of relatives (even one quite uncommon name such as Lanez), places (Maquiné Cave, the grandmother's birthplace, once visited by JS), and dates (JS death and birth dates). It also contained very private information as precise facts about the circumstances of death: "crushed by the accident," "The train from Mogi on June 8th of last year," "morning," "with friends," and "I awoke ... with a protective bandage around my head." Concerning the bandage around his head, it is worth noting that only his sister and uncle had this knowledge as only they saw his body in the mortuary and JS's wake took place with his coffin closed.

Other very intimate pieces of information were also supplied, such as "since my early days at school the idea of a train was always with me, and my preoccupation with time would make me write dates on everything" [he was a train enthusiast since childhood and used to write his name and dates on many objects and papers]. Also, "You [Dad] wanted to have a son who would work in a hospital, caring for children in need and attending to the destitute sick" [father told JS he wanted him to become a doctor to look after poor children]. Finally, a special request was made to the father who was secretly nurturing suicidal thoughts: "Do not wish to die just to be together with me" (Paraná et al., 2019).

Our research group has also investigated 13 letters from 1974 to 1979 attributed to JP, a fourth-year mechanical engineering student who drowned at the age of 24, in 1974. The first letter contained 17 verifiable items, including 3 first names, 1 surname, 1 date, names of places, and references to specific interests. Details concerning JP's circumstances of death were also given. One interesting fact was that the quality of communication evolved in the subsequent letters as they were written in a similar language to that used by JP. All those information was given despite JP's sister having stated that she only said to Chico she had lost her brother. Another interesting fact was three cases of drop-in[2] communications that were found in the

---

[2] Drop-in communicators are those alleged personalities who appear spontaneously and communicate through the medium, though unknown to any sitter present.

letters. For example, one letter presented the full name (Irineu Leite da Silva), date of death, the first names of his parents, and the cemetery where he was buried, and all this information was later verified (Rocha et al., 2014).

Studies were also conducted to investigate Chico's literary mediumistic production. Rocha (2001)—a PhD in Brazilian literature—analyzed Chico's first mediumistic poetry book *Parnassus from Beyond the Tomb*, composed of 259 poems attributed to 56 Brazilian and Portuguese poets, published in 1932 when he was just 22 years old. Rocha selected five poets and investigated common points between the poems obtained psychographically and those written by these poets when alive. Through stylistic, formal, and interpretative analysis, the conclusion suggested that the poems were not the product of mere literary imitation. For example, all 31 poems attributed to Augusto dos Anjos (1884–1914), a famous Brazilian poet, were in decasyllabic verse and were composed in a complex form specific to the alleged poet that only a specialist in linguistics would know how to assess (Rocha, 2001).

Other specific stylistic features were detected with the other poets, including personality traits and topics of interest (Rocha, 2001). In 1944, Monteiro Lobato (one of the most prominent Brazilian writers) stated that if Chico wrote that book by himself, "he could be in any academy [of letters], occupying however many chairs he likes" (Camargo, 2021, p. 92). Rocha (2008) also studied Chico's mediumistic writings attributed to the Brazilian writer Humberto de Campos, and the conclusions followed the same trend: the writings represented the style and knowledge expected of the alleged author. In this last writings, there was a variety of direct and veiled references (intertextuality) to texts published by Humberto de Campos in his lifetime, including some texts that were not publicly available when the text was automatically written (Rocha, 2008).

The alleged scientific information provided in Chico's mediumistic writings was also investigated. A study correlated many statements of several functions of the pineal gland (or epiphysis) made in books allegedly written by a deceased physician in the 1940s (when the pineal was thought to be just a vestigial gland with basically no function in adulthood) and compared with the current scientific knowledge (Lucchetti et al., 2013). For example, the writings presented clear statements for describing the pineal gland's functions of hormonal and emotional regulation such as secretes the "psychic hormones," "the function of the epiphysis in mental life is very important," and "it presides over the neural phenomena of the emotions." However, these functions attributed to the pineal gland were only reported by science many years later as only in 1958 was melatonin (the hormone secreted by the pineal gland) isolated (Lerner et al., 1958), and later it was associated with mental health (Cardinali et al., 2012; Sánchez-Barceló et al., 2010).

Finally, it is worth mentioning another type of evidence of survival produced by Chico. A forensic expert in handwriting investigated an automatically written letter and concluded that the letter contained the characteristics of the alleged author's handwriting (Perandréa, 1991). However, this kind of study needs replication.

To summarize, Chico Xavier produced a wide variety of mediumistic phenomena indicative of survival, with many markers of personal identity: memory, literary style, handwriting, features of personality, specific interests, and cultural background. These were expressed in 450 books and probably thousands of letters. It is important to note that Chico had only formal elementary education, had modest financial resources, lived before the internet age in a medium-sized city, and had to reconcile his intense mediumistic activity with his work as a public servant. He wrote letters in front of hundreds of random people who in séances that took place twice a week for decades. It is unfathomable to discern how much time, money, and energy it would have taken to build a shrewd structure of fraud so that one could obtain all the information that his letters and books contained. And, like Mrs. Piper, Chico Xavier was never caught perpetrating a fraud.[3]

Other Mediumship Studies

Many other mediums have been investigated scientifically. One such prominent but controversial medium was the Italian Eusapia Palladino (1854–1918). She was capable of a wide range of phenomena including reporting veridical information unknown to her, xenoglossy,[4] movement of objects, changes in the weight of objects, and materializations. She was assessed by many acclaimed scientists and intellectuals such as Charles Richet, Sir Oliver Lodge, Frederic Myers, Henry and Eleanor Sidgwick and Richard Hodgson (SPR members), Pierre and Marie Curie, and Henri Bergson.

The Italian psychiatrist, criminologist, and materialist Cesare Lombroso, after his own initial vehement refusal to take her seriously, investigated Palladino's sittings for years and became convinced of survival (Lombroso, 1909). Despite many negative assessments having been made about her mediumistic phenomena, based on her difficult temperament and her tendency to cheat, the results of several séances, conducted under very strict conditions and watched by rigorous experts, were impressive (Alvarado, n.d.; Bozzano, 1995; Lombroso, 1909).

Other phenomena related to mediumship and relevant to survival include the apparition of deceased personalities. There is evidence that apparitions are much more frequent around the time of death (even when the observer is not aware of the illness or the death of the perceived person), making chance or expectation unlikely explanations. In addition, there are several reports of apparitions (that convey veridical information) seen simultaneously by two or more people (Gurney et al., 1886), and very specific and accurate information unknown to any living person can be transmitted by an apparition. The case of the will of Chaffin below illustrates that.

In 1905, Chaffin wrote a will leaving all his estate to his third son. In 1919, he secretly made a new one, dividing his property equally between the four children.

---

[3] For a more detailed analysis of the survival evidence produced by Chico Xavier, see Rocha AC, Weiler M, and Casseb RF. Mediumship as the Best Evidence for the Afterlife: Francisco Candido Xavier, a White Crow. Essay submitted to the BICS competition, 2021. Available at https://www.bigelowinstitute.org/Winning_Essays/Alexandre_Roch_et_al.pdf

[4] Ability to speak a real language not normally known to the subject (Stevenson, 1974a).

This second will was handwritten and placed between two pages of an old Bible. He never mentioned the existence of this second will to anyone but left inside the pocket of an overcoat a roll of paper directing the reader to look in the Bible. In 1921, he died after a fall, and his third son inherited the property as indicated by the first will. In 1925, the second son started to have visions of his father. On one occasion, the father appeared wearing an overcoat and said, "You will find my will in my overcoat pocket" ("Case of the Will of James L. Chaffin," 1928, p. 519). Together with other witnesses, the son found the overcoat at the house of one of his brothers. In the overcoat pocket, there was a roll of paper with the father's handwriting saying to read Genesis chapter 27 of Daddy's old Bible. They found the Bible at their mother's house, and the second will was folded on that page. The court adjudged the new will to be valid. Moreover, the analysis of ten witnesses validated Chaffin's handwriting ("Case of the Will of James L. Chaffin," 1928).

While most of these previous studies were in-depth, qualitative investigations conducted with exceptionally gifted mediums, more recently, there has been a prevalence of quantitative studies assessing the accuracy of mediumistic information involving several mediums under strict conditions. They often use blind protocols, where the medium is blinded to information about the sitter and the discarnate personality. In addition, a "proxy sitter" is used (i.e., the person who has contact with the medium is not the actual sitter), so cold-reading and fraud can be excluded as explanations. Sitters are blinded to the readings (they do not know which reading is intended for them), so they are not emotionally biased to accept the reading, and the researchers are also blinded. One researcher is designated to interact with the mediums, while another interacts with the sitters to minimize alternative explanations for the phenomenon, such as telepathy, fraud, and cues from sitters and experimenters.

Beischel and Schwartz (2007), for example, assessed eight mediums and eight sitters. The mediums only received the first name of the discarnate personality, and the sitters received two readings—the correct one and a control, in which 81% of the sitters chose the correct readings (Beischel & Schwartz, 2007). A larger study including 20 mediums and 86 readings successfully replicated these findings (Beischel et al., 2015).

In order to assess the accuracy of the information provided by the mediums, a recent meta-analysis[5] included a total of 18 controlled, blind experiments, conducted between 2001 and 2019, that quantitatively assessed the accuracy of anomalous information received by mediums. The scores for the intended readings were statistically higher than for the control readings, with an effect 0.18 above the level of chance (Sarraf et al., 2021).

Neurophysiological findings under mediumistic states have also been investigated. Our group studied brain activation using a *single photon emission computed tomography* (SPECT) of ten mediums during automatic writing (trance) and regular writing (usual state of consciousness) and compared the complexity of the texts

---

[5] In health sciences, the most comprehensive procedure for summarizing scientific evidence, and the one often considered to be the gold standard, is the performance of systematic reviews with meta-analysis of data provided by quantitative studies.

produced in both conditions. One may think that the more complex a text is, the greater the activation of brain areas associated with creativity and planning. However, the opposite was the case. During the automatic writing, mediums produced more complex texts despite showing lower brain activation compared to the control writing (Peres et al., 2012).

This lower brain activation during the process of automatic writing is in line with mediums' claims that, in trance states, they report mind content that does not come from their own thought process. Other neurophysiological studies found that a medium in trance who presents reduced frontal brain activity in the electroencephalogram (EEG) produced more accurate information (Delorme et al., 2013) and that mediums experiencing contact with an alleged dead personality produced different brain activation patterns when compared to imagining this contact (Mainieri et al., 2017).

Alternative Explanations

The pioneers of modern mediumship research were always aware of the constant presence of fraud and superstition in these phenomena. By the mid-nineteenth-century, investigators of mediumship were already developing methods to assess when mediums were lying or faking. Allan Kardec discussed fraudulent practices in his 1861 *The Mediums' Book* and concluded that one of the best ways to minimize fraud was to eliminate the motivation to perform it: i.e., financial profit (Kardec, 1861). SPR members developed several strategies to control for fraud: being witnesses to séances themselves, interrogation of sitters and mediums, detailed contemporary séance transcripts were written and signed by witnesses (Johnson, 1980, p. 100), and hiring detectives to shadow mediums. Studies conducted today use a very strict protocol in which sitters, mediums, and researchers are all blinded to each other. So, although fraud may explain away several cases, there is a consensus among qualified researchers that it cannot explain away the best-controlled ones.

The emotional fragility of bereaved sitters might bias them to over-endorse the readings of mediums. However, a recent study our group conducted did not support this hypothesis. Despite almost 90% of the bereaved sitters believing in mediumship and wishing to receive a letter, they did not endorse letters with generic content (Freire et al., 2022). In addition, research protocols are designed to deal with this bias. For example, the sitters blindly receive the intended reading and a few control ones, so they score the readings without knowing which one is intended for them (Beischel & Schwartz, 2007). Also, many statements are specific such as "You [Dad] wanted to have a son who would work in a hospital, caring for children in need and attending to the destitute sick" in the letter written by Chico (Paraná et al., 2019).

Another alternative hypothesis is coincidence, the chance that part of the information reported by the medium matches someone's life. Most modern studies use statistical methods precisely to control for chance. For example, when sitters blindly scored the intended readings higher than the controls with a p value of 0.007, it means that this difference has a 0.7% chance of having been found at random[6] (Beischel & Schwartz, 2007). The meta-analysis of 18 controlled studies confirmed that the pool of positive results is very unlikely to have happened by chance (Sarraf et al., 2021). Moreover, it is very hard to attribute to chance very specific and personal information and features such as those provided by Chico Xavier and Mrs. Piper such as uncommon personal names, life occurrences, and literary styles.

Mediumship is a complex phenomenon, and well-designed recent studies, as well as reports of challenging and complex cases such as those of Chico Xavier and Mrs. Piper, confirm that mediums can obtain veridical information and demonstrate nonverbal skills beyond the explanatory capacity of conventional alternative hypotheses. Mediums such as Chico Xavier and Mrs. Piper produced veridical information under strict conditions and demonstrated personality traits and skills consistent with the alleged deceased and beyond their acquired abilities and knowledge. Taking the evidence as a whole, survival of consciousness emerges as the most reasonable and empirically supported explanatory hypothesis.

Near-Death and Out-of-Body Experiences

Near-death experiences (NDEs) are unusual or extraordinary experiences, often transformative, vivid, and with a transcendent aspect, that occur in life-threatening conditions (Holden et al., 2009; van Lommel, 2013). Individuals who experience NDE usually report similar elements in their experiences, such as out-of-body experience (OBE), i.e., a clear perception of leaving the body, often seeing it from an external vantage point; seeing a tunnel or a bright light; feelings of peace; meeting deceased relatives or beings of light; life review; and the sensation of returning into the body (Charland-Verville et al., 2020; Martial et al., 2017; Sabom & Kreutiziger, 1977; van Lommel, 2013). They often feel deeply transformed by the experience, giving new meaning and adding different insights to their lives. Commonly, fear of death decreases, and belief in an afterlife increases (van Lommel et al., 2001).

NDE descriptions and aftereffects seem to be similar across different countries, cultures, and epochs (Shushan, 2009, Ch. 3). One study found strong similarities in 15 different phenomenological features among NDEs reported before and after 1975 (when NDE became popular with the publication of the book "Life After Life," Moody Jr) (Athappilly et al., 2006).

A review of prospective studies found a 17% incidence of NDE among those in life-threatening events (Zingrone and Alvarado in Holden et al., 2009). OBEs and NDEs have been described in different cultures and contexts throughout history (Bozzano, 1997; Shushan, 2009), indicating *similar empirical occurrences* beyond

---

[6] Usually in science, chance below 5% (or, in stricter conditions, 1%) is considered "statistically significant," i.e., unlikely to have happened by chance.

cultural factors. NDEs seem to be a more robust and specific variation of the broader concept of OBEs, and both challenge materialistic explanations. Possibly, NDE provides more evidence for mind-body dualism than for survival. However, since it is strong evidence for dualism, it makes the survival of bodily death more plausible. Indeed, historically, NDEs have been linked to a belief in an afterlife (Shushan, 2009).

A somewhat related phenomenon is "terminal lucidity," the unexpected recovery of mental function just before death in patients suffering from severely compromised brains (e.g., advanced dementia) (Nahm et al., 2012). It resembles the vivid experiences NDE experiencers have when their brains function are severely impaired, suggesting that physical damage to the brain may not impair consciousness.

Veridical Perceptions

The most relevant feature of NDE as evidence for survival is the report of OBE with veridical perception of events occurring during cardiac arrest when the brain is not functional. One of the most impressive cases is a 35-year-old American woman subjected to neurosurgery for a brain aneurysm. During this rare, high-risk procedure, her body was cooled to 60.8 °F (16 °C), her heart was stopped, and the blood from her brain was drained. Her body and brain were extensively monitored, she wore earphones, and her eyes were taped. In hypothermic cardiac arrest, the electroencephalographic response was a "flatline" (i.e., no brain activity could be detected).

Despite these strict conditions, she described a profound NDE during which she reported veridical facts such as a detailed description of the bone saw and a female voice saying her veins and arteries were too small (Kean, 2017, Ch. 8; Sabom, 1998). Also, she described that the earphones emitted a natural D note, which was subsequently confirmed. The neurosurgeon who operated on her stated:

> I don't think the observations she made were based on what she experienced when she went into the operating theatre. They were just not available to her. For example, the drill and so on were covered up and invisible. They were inside their packages. The packages in which the drill and other equipment were stored would not have been opened before the operation, and in any event, Reynolds's eyes were taped shut. ... At that stage of the operation nobody can observe or hear in that state. I find it inconceivable ... that there was any way for her to hear those through the normal auditory pathways. (Smit, 2008, p. 308-309)

Mr. Laufmann is another example who reported a veridical perception of an acquaintance of him during an NDE (Muldoon & Carrington, 1984, p. 83):

> I was standing there in the middle of the room and distinctly saw my dead body lying upon the bed.... I started to leave the room and met one of the physicians, and was surprised that he said nothing to me, but since he made no effort to stop me, I walked out into the street where I met an acquaintance of mine, Mr. Milton Blose. I tried to greet Mr. Blose by hitting him on the back, but my arm went through him.... It was impossible for me to attract his attention.... I saw that he went across the street and looked into a shop window where a miniature 'Ferris wheel' was on display.

Mr. Blose left a testimonial letter verifying that Mr. Laufmann's report about him was true (Muldoon & Carrington, 1984, p. 84).

Another impressive case of veridical perception during an NDE was published in The Lancet, a leading global medical journal. A nurse reported the case below of a man in cardiac arrest with no blood circulation and pupils unresponsive to light (van Lommel et al., 2001; Rivas et al., 2016):

> After admission, he receives artificial respiration without intubation, while heart massage and defibrillation are also applied. When we want to intubate the patient, he turns out to have dentures in his mouth. I remove these upper dentures and put them onto the 'crash cart'. Meanwhile, we continue extensive CPR [cardiopulmonary resuscitation]. ... Only after more than a week do I meet again with the patient... The moment he sees me he says: 'Oh, that nurse knows where my dentures are'. I am very surprised. Then he elucidates: 'Yes, you were there when I was brought into hospital and you took my dentures out of my mouth and put them onto that cart, it had all these bottles on it and there was this sliding drawer underneath and there you put my teeth.' When I asked further, it appeared the man had seen himself lying in bed, that he had perceived from above how nurses and doctors had been busy with CPR. He was also able to describe correctly and in detail the small room in which he had been resuscitated as well as the appearance of those present like myself. (van Lommel et al., 2001, p. 18-19)

There are dozens of reports in the literature of cases similar to those mentioned above (Moreira-Almeida & Santos, 2012; Rivas et al., 2016) including in blind individuals (some blind from birth) who also describe detailed visual representations and veridical perceptions similar to those of sighted persons (Ring & Cooper, 1997). In addition, one cardiologist found that NDE experiencers reported resuscitation procedures (they claimed to have observed during OBE) more accurately than patients who were resuscitated but did not report NDE (Sabom, 1982). Lastly, a review of 93 reports of potentially verifiable out-of-body perceptions during NDE found that about 90% were completely accurate, 8% contained some minor error, and only 2% were utterly erroneous (Holden et al., 2009).

As far as OBE is concerned, one experiment investigated, for four nights in a sleep laboratory, a Miss Z, who used to have spontaneous OBE at night (Tart, 1968). After she fell asleep, the researcher placed a random five-digit number on a high shelf above Miss Z's headboard. As Miss Z used to wake up shortly after the OBE, the researcher verified that the OBE occurred concomitantly with an unusual EEG pattern, different from the waking or sleeping patterns. On the fourth night, Miss Z awoke and reported having seen the target number and correctly said "25,132" while she was continuously monitored, not being able to move to see the number without disconnecting the EEG wires (Tart, 1968).

NDE experiencers often report seeing deceased people; some of them they did not know were dead at the time of the NDE. In a sample of 665 NDEs, 21% reported having encountered a deceased person while only 4% a living one (Greyson, 2010). This experience of seeing deceased people is more frequent in those NDE experiencers who were closer to death (Kelly, 2001). Van Lommel (2013) reported an interesting case in which a man claimed to have seen a man he did not know and that, years later, he found out that the man was his biological father:

> During my cardiac arrest I had an extensive experience (…) and later I saw, apart from my deceased grandmother, a man who had looked at me lovingly, but whom I did not know. More than 10 years later, at my mother's deathbed, she confessed to me that I had been born out of an extramarital relationship, my father being a Jewish man who had been deported and killed during the Second World War, and my mother showed me his picture. The unknown man that I had seen more than 10 years before during my NDE turned out to be my biological father. (van Lommel, 2013, p. 21)

NDE experiencers' report of perceiving NDE memories as true and even more real than the facts of ordinary life is supported by studies. A study with 122 NDE experiencers in the USA found that their NDE memories contained more features of real memories (e.g., perceptual information [color and sound], contextual information [surrounding time and place], and meaningful details such as emotion) than those related to real or imagined events (Moore & Greyson, 2017). Similar results were also reported in Belgium (Thonnard et al., 2013) and Italy (Palmieri et al., 2014). There is also evidence that NDE experiencers do not embellish their reports over time (Alvarado & Zingrone, 1998), and a study with 72 NDE experiencers found that the details of their NDE reports did not change even after two decades (Greyson, 2007).

Alternative Explanations

The most common arguments against the hypothesis of survival relating to NDE cases include psychological or cognitive processes such as imagination or fear of death, cultural expectations, mental symptoms such as hallucination, and medical conditions such as drug use, oxygen deficiency, or neurohormonal dysregulation. However, most scientists who conducted the best and largest studies with NDE experiencers agree that these conventional explanations cannot fully explain NDE, which seems to be an authentic experience that requires an explanation beyond psychology and physiology (van Lommel, 2011, p.19). The main arguments are:

- Clear consciousness and sharp mental function while the brain is electrically inactive or in conditions of severely impaired brain function (Moore & Greyson, 2017; Parnia, 2007; van Lommel, 2011).
- NDE experiencers consistently report enhanced consciousness, lucid and vivid experiences perceived to be more real than reality itself, and measures to assess memory quality support these claims (Cook et al., 1998).
- Deep and lasting impact on a person's values and beliefs, which rarely happens after a state of confusion in critical patients (van Lommel et al., 2001; Trent-Von Haesler & Beauregard, 2013).
- When we imagine ourselves, we do not usually see our physical body from a different position in space, unlike the reports in cases of NDE.
- Communication during NDE is often reported as thorough thought transference which is an unusual experience in everyday life (Cook et al., 1998).

- Core NDE features, that have been reported throughout the ages and cultures, did not change after NDE became popular and often run counter to patients' previous religious or materialistic beliefs (Greyson, 2007, 2015; Holden et al., 2009; Rivas et al., 2016).
- Among patients who survived cardiac arrest, those who had an NDE are no different from those who did not when it comes to their religious beliefs, educational level, fear of death, duration of cardiac arrest, and medication used (Greyson, 2007; van Lommel et al., 2001; Trent-Von Haesler & Beauregard, 2013).
- NDEs cannot be explained by hypoxia (or other factors related to brain injury) since NDE can also occur when the brain is intact, such as in a severe car accident where the person is uninjured.
- Verified veridical perceptions of facts that happened when the brain is not functional, especially during cardiac arrest, since cortical electrical brain activity ceases 10 to 30 seconds after circulatory arrest. This excludes the hypotheses that the person may be listening through his ears, seeing through his eyes, and storing the experience in his brain or that the memory of the NDE may be created when consciousness is either being lost or regained, rather than in the period of clinical death itself (Holden et al., 2009; Parnia, 2007; Rivas et al., 2016; Sabom, 1998).
- Hallucinatory experiences often happen during hypoxic or confusional states in critical patients. However, these experiences are usually fragmentary, frightening, with clouded consciousness, and, after recovery, the patients do not remember them or acknowledge them as unreal, caused by physical disturbances. NDE differs from all these features (Parnia, 2007).

Reincarnation

Cases of the Reincarnation Type (CORTs) are those in which individuals claim to remember a previous existence. These cases have been reported in different countries from East to West and involve children usually between 2 and 4 years old, who spontaneously start talking about a previous life providing details such as names and places (Moraes et al., 2022; Stevenson, 1983). Sometimes, these very young children also exhibit certain behaviors associated with the previous personality and unrelated to their current lives (Moraes et al., 2022; Stevenson, 1983).

The idea of reincarnation is one of the most universal among the spiritual concepts of humanity. Numerous tribes and peoples in the Americas, Oceania, Sub-Saharan Africa, Siberia, Indonesia, and Europe believe in reincarnation (Obst, 2009). Also, historical and archaeological research has concluded that many afterlife beliefs and burial rituals in primitive societies actually denoted some sort of reincarnation but were previously interpreted as a different form of belief in the afterlife (Farnell, 1921, p. 356; Helck & Otto, 1975, Vol. III; Schmidt-Leukel & Bauer, 1996; Shushan, 2009). Among the largest contemporary world religions, Hinduism and Buddhism regard reincarnation as a core principle. Nor is it an alien concept in the history of Judaism, Christianity, and Islam (Bulğen, 2018; Fadiman & Frager, 1997; MacGregor, 1982).

One of the greatest Jewish historians, Flavius Josephus, and the best-known Jewish philosopher of antiquity, Philo of Alexandria, supported reincarnation. Josephus also suggests that some Pharisees believed in reincarnation (Josephus, 1900, p. 210–214). As Christianity and Islam derived from or are heavily influenced by Judaism, it is not surprising that reincarnation might be associated with these religions too. In religion, however, it is common to observe a considerable gulf between the "perceived doctrine" held by most priests and the position of the scholars.

Among Christians, the idea of reincarnation has never been too distant, both because of Greek and Celtic influences and because of passages in the New Testament that may be interpreted as referring to reincarnation. Helmut Obst (2009), one of the many Lutheran intellectuals dedicated to monitoring the idea of reincarnation in the West, offered an in-depth analysis of the passages where Jesus tells Nicodemus that a man should be reborn and when the apostles inferred that Elias could be reincarnated as John the Baptist. The Anglican theologian Geddes MacGregor (1982) came to a similar conclusion, stressing that Christian doctrine is in no way incompatible with reincarnation.

Although rejected by most Muslims, belief in reincarnation does occur among the Islamic population and is also supported by some Muslim intellectuals as being fully compatible with the Koran (Bulğen, 2018).

The key defenders of reincarnation in modern-day western societies are G. E. Lessing and Allan Kardec (Obst, 2009), exerting a strong influence upon a number of other authors, probably mainly due to the great explanatory power of their arguments, which connect morality, metaphysics, natural theology, and anthropology.

In short, the concept of reincarnation is astonishingly widespread, almost as much as the survival belief itself among more primitive cultures, and this universality reveals a cognitive disposition to be explained by rigorous research. Still, nowadays, even in majorly Christian Western countries, reincarnation is a belief held by a large proportion of the population (27% in Western Europe, 33% in the USA, and 37% in Brazil) (DataFolha, 2007; Inglehart and Basanez, 2004; Gallup, 2003).

One main opponent of the idea of immortality, who nevertheless considered the preexistence of the soul, was the philosopher David Hume. In his essay on Immortality, he argues that the concept of one single life that leads to immortality would contradict what we know about nature (regularity). If, however, souls were substances that could survive physical death, they should also logically preexist birth, which would make the concept of reincarnation the most logical of options among survivalist hypotheses (Hume, 1849, p. 226).

McTaggart offered a purely philosophical argument in support of reincarnation. According to him, there is less consistency in admitting the afterlife without the soul's preexistence than with it. Moreover, experience also suggests that some human traits may be better explained by reincarnation than by any other theory.

Such is the case with prodigious children (McTaggart, 1930, p. 122–124), a profound sense of purpose or mission in young people, and deep love (Geach, 1979, p. 165–167). As with many others, McTaggart sees existential reasons in support of reincarnation. People always seek to finish what they started, execute what they planned, and maintain their feelings for loved ones more frequently than they change them. Should we not expect that our dearest plans and projects, our beloved ones, would remain a significant concern in the afterlife?

To the surprise of some, German atheist philosopher Georg Lichtenberg had developed almost the same argument as McTaggart's more than a century earlier (Kleisner, 1998, p. 147–148). According to Lichtenberg, who despised religion as superstitious, reincarnation should be respected as a theory more fitting to the scientific understanding of the world, based on transformation instead of creation out of nothing. Additionally, Lichtenberg observed that talents and inclinations had to be explained by higher psychological functions. It would be nearly impossible to explain familiarity with complex ideas and skills or sentiments, that take a long time to grow, without the supposition of previous existence and the longevity of the mind.

The poet and naturalist Goethe played a significant role in the debate over survival and reincarnation among the early nineteenth-century German thinkers. More than once, he claimed to have hints of past lives memories and "affinities" but believed that the idea of reincarnation emerged out of the need to strike a balance between science, natural religion, and art (Coelho, 2012; Scholz, 1934).

In Obst's conclusion about the relevance to our times of the idea of reincarnation, he stresses the fact that it is a common belief among a wide spectrum of religions, atheists, and agnostics, so that the empirical support for it and the flexibility, i.e., broad acceptability of the concept, make it fit for a general fissure in the materialistic mentality (Obst, 2009, p. 265).

Since the influential research of Ian Stevenson, however, the evidential approach to the idea of reincarnation has prevailed, to the point that some philosophers already claim that it would be irrational to deny that the evidence leads to the conclusion that reincarnation is a fact of nature (Almeder, 1992, p. 62).

Cases of Reincarnation Type

More than a century ago, researchers such as Gabriel Delanne and Albert de Rochas began systematic studies of alleged, spontaneous memories of past lives (de Rochas, 1911). In the mid-twentieth century, Ian Stevenson, then chairman of the Department of Psychiatry at the University of Virginia, began systematically investigating CORT. In 1960, he published a review of 44 published CORTs (Stevenson, 1961). Stevenson then started hearing about more cases and looking for new ones while on his travels, initially to India and Sri Lanka. Soon, he realized that the phenomenon was much more common than anyone had suspected. As a result, he dedicated more than 50 years to the scientific investigation of this subject in different cultures

(Tucker, 2008). The British Medical Journal published Stevenson's obituary entitled "Psychiatrist who researched reincarnation with scientific rigour" (Tanne, 2007).

Researchers from different cultures and academic backgrounds have replicated Stevenson's CORT findings at different times and in diverse populations (Moraes et al., 2022). More than 2500 cases of young children who claimed a past life have been recorded in the files of the Division of Personality Studies files at the University of Virginia (KEIL & Tucker, 2005).

Scientific studies of CORT usually describe in detail the findings of each case. The examples below illustrate the kind of information children provided:

1. A boy from Thailand, Chanai Choomalaiwong, started claiming he had been a teacher named Bua Kai and provided a variety of details about his life: he had been killed by a gunshot while riding a bike to school; his parents,' wife's, and children's names; and the name of the village where he had lived. He would often insist that his grandmother take him to the village. When he was 3 years old, his grandmother took him there (a place some 12 miles from where they lived). Chanai led his grandmother to a house where the parents of Bua Kai, whose story was consistent with that described by Chanai, lived. There, Chanai recognized Bua Kai's parents among other family members. He also recognized other family members as well as objects that belonged to Bua Kai. Moreover, Chanai was born with two birthmarks that matched the gunshot wound that killed Bua Kai (Tucker, 2008).
2. The American boy Sam Taylor, when he was 18 months old, said to his father while he was changing his diaper "when I was your age, I used to change your diapers" (Tucker, 2008, p. 246). After that, he began saying he had been his grandfather in the previous life and provided more details of his and family members' lives. Once, Sam's mother asked him if he had had brothers or sisters in the previous life, and he correctly and spontaneously stated that he had a sister who had been murdered. When Sam was 4.5 years old, his mother showed him a picture of a class including 16 boys and 11 girls, and Sam, with no difficulty, chose the young man corresponding to his grandfather. Sam provided other correct information such as grandfather's favorite place in the house, the food his grandmother gave him every day (milkshake) and what appliance she prepared it with (a processor, and not a blender), and that Sam's father had a toy steering wheel as a child. He also spontaneously showed his grandfather's car in a photo (Tucker, 2006, Ch. 7).

These examples illustrate the specific nature of the information reported by the children, and many cases are like Chanai's: they often involve families who lived far from each other and did not know each other. So, it is implausible that the child had access to this information, especially considering the familiarity with which they express the knowledge about the life of the alleged previous personality. Moreover, it is important to note that these children describe ordinary past lives, despite being in a psychological development phase in which they are prone to fable. If these accounts were merely the result of childhood fantasies, children would tend to produce claims of more fantastic and extraordinary lives.

In some cases, the researchers arrive before the statements provided by the children have been verified by relatives. The boy KA in Hatay, Turkey, among several other details said that his previous family's name was Karakas, he had been a rich Armenian Christian, he lived in a large three-story house near the water in Istanbul, his wife and children had Greek first names, he lived near a well-known woman named Aysegul who was an art dealer, and he had been killed and his wife was involved. KA's parents had a high level of education and did not attach importance to the case. The investigators were able to verify that 15 statements were fully correct, 3 were partially correct, and 4 could not be verified (KEIL & Tucker, 2005).

Furthermore, three leading reincarnation researchers from three locations (the USA, Germany, and Tasmania) replicated Stevenson's findings in a sample of 123 CORTs in 5 different countries (Burma, Thailand, Turkey, Sri Lanka, and India). The main characteristics of their sample were similar to those described by Stevenson. They also described in detail three cases that, in total, contained 67 statements. Of these, 49 (73.1%) were truthful information that could be verified, 11 (16.4%) could not be verified, and 7 (10.4%) were incorrect. This incorrect information was present in the case of just one individual, who made 28 statements, 17 of which provided specific, correct information (Mills et al., 1994).

Other common features verified in CORT are personality traits, unusual behavior, phobias or symptoms of posttraumatic stress disorder, skills not learned in life, and birthmarks/birth defects (Keil & Stevenson, 1999; Stevenson, 1977b, 1990; Tucker, 2008). Almost a quarter of children who claim to remember previous lives engaged in play that was not germane to the current life but rather consistent with the claimed memories.

One case, for example, is related to a child who played at managing a nightclub. "He arranged boxes to represent a bar and placed bottles on them. He assigned roles in the club to neighborhood girls and gave one a stick that represented the microphone held by a singer. He put out two chairs for the wives of the club's owner" (Stevenson, 2000, p. 562). Naming toys with some previous personality's children or relatives' names is reported too (Stevenson, 2000).

Another interesting case concerning specific behaviors includes children in Myanmar who claimed to remember a previous life as Japanese soldiers killed in Burma (now Myanmar) during World War II. These children showed typical behavior of the Japanese, despite the oppressive history of the Japanese Army in Burma (so the contact between Burmese villagers and Japanese soldiers was minimal), and most of these children were born after the Japanese occupation had ceased (Keil & Tucker, 2000; Stevenson, 1983, 1997, 2001). Those children reported a desire to return to Japan—a preference for Japanese clothing, raw or partially cooked fish, raw eggs, sweet foods, and strong tea—and complained that Burmese food was too spicy, a resistance to learning Burmese. Also, they spoke with a "foreign" accent and sat in an unusual way—on the floor with buttocks resting on heels (Stevenson & Keil, 2005).

Other behaviors that appear connected to the personality's previous life are inexplicable phobias that are appropriate to the life of the alleged previous personality. An example is the case of Ranbir Yadav, who, at 2.5 years old, said, among several other things, his name was Sohan and that he was hit by a bus while riding a bike. Ranbir was afraid of buses and bicycles and refused to ride on buses. Further investigations verified as accurate his claims about the previous personality who died as he had described (Pasricha, 1996). In many cases, the children manifested phobias before they could speak, such as a baby in Sri Lanka who so struggled with water (and later reported having drowned in a previous life, which was verified) that it took three adults to bathe him (Stevenson, 1990).

These phobias usually start between the ages of 2 and 5. Systematic inquiries revealed that these phobias were not associated with personal traumatic experiences or models presented in the environment, which are common explanations for phobias (Pasricha, 1996). Phobias occur more frequently among children who claim to remember dying violently when compared to those who remember natural deaths. Although past-life memories usually fade away as children grow older, phobic behavior may persist (Pasricha, 1996; Tucker, 2008).

Surprisingly, about 35% of CORTs have birthmarks or birth defects that match wounds suffered in the alleged previous life. The birthmarks are usually skin marks—such as hypopigmented or hyperpigmented areas, or, sometimes, a severe birth defect has been reported. Stevenson reported 49 cases in which he obtained medical records relating to the injuries of the alleged past-life personality. Of those, 43 (88%) demonstrated concordance between the body location and the features of the birthmark and the wound suffered in a previous life. Some children have two or more concordant birthmarks, making the cases less prone to being at random (Stevenson, 1993, 1997). Below some examples are described:

1. A Thai man had a large linear verrucous epidermal nevus (a wartlike, reddish or brown lesion) on the back of the head, and when he was a child, he claimed he remembered the life of his uncle who had been struck on the same region of the head with a heavy knife (Stevenson, 1993).
2. Chanai, the Thai boy mentioned at the beginning of this chapter, was born with two birthmarks (Tucker, 2008). The small, round one on the back of his head was consistent with the gunshot entry wound on the back of Bua Kai's head and the larger, more irregular one on the front, above his left eye, with the gunshot exit wound on Bua Kai's forehead (Tucker, 2008, p. 245). Scholars in forensic ballistics know that the bullet entry wound is smaller and more regular than the exit one.
3. Purnima is a girl from Sri Lanka who claimed, since she was 3 years old, that in her previous life she lived as a man in a family of incense makers. She made a variety of verified statements: where the previous family worked [the incense factory was near a brick factory and a pond] and lived [on the other side of the river from the Kelaniya Temple], the incense brand names [Ambiga and Geta Pichcha], had had two wives and two younger brothers, gave mother's name [Simona], and had died in a traffic accident. She had a sizeable hypopigmented birthmark to the

left of the midline of her chest and above her lower ribs that matches the autopsy of Purnima's alleged previous personality (Haraldsson, 2001).

Purnima even described how incense was made: "There are two ways to make it. One uses cow dung, the other is from ash from firewood (charcoal). A paste is made, and then a thin stick is cut from bamboo, and some gum is applied to the bamboo stick. Then the stick is rolled over the paste, and then something is applied to obtain a nice smell." "What is charcoal made from and how is it produced?" asked the researcher, and she answered: "When firewood is burnt you get charcoal." Purnima's father only knew that incense was made from cow dung; the mother and the researcher's interpreter knew nothing about it (Haraldsson, 2001 p. 21).

Alternative Explanations

One counterargument to the survival hypothesis for CORT is that it is a cultural phenomenon since CORTs are more commonly reported in cultures where reincarnation beliefs are prevalent. However, CORTs have been reported in different cultures, including those in which the belief in reincarnation is less prevalent, as in Europe and America (Stevenson, 1974a, 1983). The belief can create a validating environment for the manifestation of the phenomenon, but it is not imperative for the emergence of cases. According to Tucker, "cases have been found wherever anyone has looked for them, including all continents except Antarctica" (Tucker, 2008).

Moreover, the cases have significant similarities across different cultures: age of onset of the communication about the alleged previous life and the age when this ceases, high prevalence of alleged violent deaths, the average number of statements made by each child, and the presence of unusual behavior (Stevenson, 1983). Similarities have also emerged over the course of time (Pasricha & Stevenson, 1987). For example, two different groups of cases that occurred about one generation apart were similar with regard to the presence of birthmarks and birth defects, the mean age of first memory claim, claims about the previous mode of death, phobias, mention of alleged previous names, high prevalence of alleged violent death, and the interval between alleged previous life and current life (Keil & Stevenson, 1999).

The family's wishes could also influence a child to fabricate alleged memories from past lives. However, it is important to remember that very young children spontaneously start making these statements, typically about very ordinary lives. Usually, there is no detectable material gain or desire for prestige associated with these memories. Parents most often do not stimulate these claims. A study reported that the initial attitude of mothers ranged from encouragement (21%) to neutral or tolerance (51%), discouragement (28%), and later changed from being neutral to taking measures to suppress the memories (Pasricha, 2011). Indian children, for example, often claim past lives in inferior castes, as in the case of Kumkum, whose father was a landowner and physician and never allowed Kumkum to visit the previous alleged family as "he was not proud that his daughter seemed to remember the life of a blacksmith's wife" (Stevenson, 1974b; Tucker, 2008).

Another possible hypothesis is that CORT children would have psychological problems and are prone to fantasies. None of the three different studies comparing psychological tests in CORT children with controls found differences in suggestibility or tendency to confabulate (Haraldsson, 1995, 2003). CORT children are usually healthy and lead normal lives. Two of these studies reported that they showed a higher level of cognitive function, better performance in school, and better memory than their peers (Haraldsson, 1995, 1997).

Unfortunately, many CORTs cannot be verified, i.e., children make statements, but a deceased person corresponding to the statements could not be found. These cases are called "unsolved" cases. Hence, one may argue that many children might talk about previous lives and that only a minority matches a dead person by chance. However, unsolved cases are the minority. For example, of 856 CORT cases in 6 different cultures, only 33% were unsolved (Cook et al., 1983). Moreover, unsolved cases present similar patterns to solved cases: age of start, mentioning the mode of death, the high rate of alleged violent ways of death, and phobias reported (Cook et al., 1983).

Some differences between solved and unsolved cases, however, might explain why some cases are unsolved: less frequently these children reported the previous personality's name and violent death (so less traumatic memory), and they stopped talking about the previous life earlier (Cook et al., 1983). So, these data are consistent with the hypothesis that unsolved cases are weaker forms of actual cases of reincarnation that, because of unknown factors, could not reproduce the verifiable features of solved cases.

Finally, one may argue that, after the families meet each other, they would embellish or give a new interpretation to wrong statements and bias the interpretation to fit the alleged person's life. If this is true, the percentage of accurate information should be higher in those cases where the family met each other before any record could be made, compared to those whose records were made afterward. Contrary to this hypothesis, no difference was found in the percentage of correct statements between both kinds of cases (Schouten & Stevenson, 1998). In addition, most CORTs cannot be explained by the unproven theory of inherited memory since most cases occur in different families (Stevenson, 1974a), some even in different countries.

To summarize, the idea of reincarnation is one of the most ancient and universal among different spiritual concepts of humanity. CORT started being scientifically investigated in the twentieth century by Ian Stevenson, a psychiatrist recognized for the rigor of his scientific method. Thereafter, several scientists around the globe (Moraes et al., 2022) have replicated these findings, creating a robust body of thousands of similar cases of very young children claiming to remember past-life memories in different times and cultures. The reincarnation hypothesis is unique for its explanatory value as it can explain a number of different events (Stevenson, 1977a):

- Accurate memories recorded before verification by the family (Schouten & Stevenson, 1998; Stevenson & Samararatne, 1988).
- Birthmarks and birth defects consistent with injuries suffered by the alleged previous personality (Stevenson, 1993, 1997).
- Emotional reactions to specific trigger and posttraumatic symptoms coinciding with the mode of death of the alleged previous personality.
- Skills such as Purnima's knowledge of making incense (Haraldsson, 2001).
- Findings which have been replicated by different researchers and in different cultures.

References

Alighieri, D. (1993). *Vida Nova*. Guimarães Editora.
Almeder, R. F. (1992). *Death and personal survival: The evidence for life after death*. Rowman & Littlefield Publishers.
Alvarado, C. S. (n.d.). *Eusapia Palladino | Psi Encyclopedia*. Retrieved August 15, 2021, from https://psi-encyclopedia.spr.ac.uk/articles/eusapia-palladino#Literature
Alvarado, C. S., & Zingrone, N. L. (1998). Factors related to the depth of near-death experiences: Testing the "embellishment over time" hypothesis. *Imagination, Cognition and Personality, 17*(4), 339–344. https://doi.org/10.2190/VYQP-867C-NEWQ-F054
Athappilly, G. K., Greyson, B., & Stevenson, I. (2006). Do prevailing societal models influence reports of near-death experiences?: A comparison of accounts reported before and after 1975. *Journal of Nervous & Mental Disease, 194*(3), 218–222. https://doi.org/10.1097/01.nmd.0000202513.65079.1e
Beischel, J., Boccuzzi, M., Biuso, M., & Rock, A. J. (2015). Anomalous information reception by research mediums under blinded conditions II: Replication and extension. *Explore, 11*(2), 136–142. https://doi.org/10.1016/j.explore.2015.01.001
Beischel, J., & Schwartz, G. E. (2007). Anomalous information reception by research mediums demonstrated using a novel triple-blind protocol. *Explore (New York, N.Y.), 3*(1), 23–27. https://doi.org/10.1016/j.explore.2006.10.004
Bigg, C. (1886). *The Christian Platonists of Alexandria*. At The Clarendon Press.
Bozzano, E. (1995). *Animismo ou Espiritismo? Qual dos dois explica o conjunto dos fatos? [Animism or spiritism? Which of the two expalins the ensemble of facts?]*. Federação Espírita Brasileira.
Bozzano, E. (1997). *Povos primitivos e manifestações supranormais*. Editora Jornalistica FE.
Bruno, G., de Lucca, R., Blackwell, R. J., Bruno, G., & Bruno, G. (1998). *Cause, principle, and unity*. Cambridge University Press.
Bulğen, M. (2018). *Reincarnation (Tanāsukh) According to Islam: Compara-tive, Historical and Contemporary Analyses [İslam Dini Açısından Reenkarnasyon (Tenâsüh): Tarihi ve Günümüz Açısından Bir Karşılaştırma]*. https://doi.org/10.5281/ZENODO.1488657.
Camargo, L. (2021). Diálogos incomuns: Monteiro Lobato e o Espiritismo. *Brasil/Brazil, 34*(64), 85–98.
Cardinali, D. P., Srinivasan, V., Brzezinski, A., & Brown, G. M. (2012). Melatonin and its analogs in insomnia and depression. *Journal of Pineal Research, 52*(4), 365–375. https://doi.org/10.1111/j.1600-079X.2011.00962.x
Case of the will of James L. Chaffin. (1928). *Proceedings of the Society for Psychical Research, 36*, 517–524.
Charland-Verville, V., Ribeiro de Paula, D., Martial, C., Cassol, H., Antonopoulos, G., Chronik, B. A., Soddu, A., & Laureys, S. (2020). Characterization of near death experiences using text

References

mining analyses: A preliminary study. *PLoS One, 15*(1), e0227402. https://doi.org/10.1371/journal.pone.0227402

Chibeni, S. (2020). Spiritism: An experimental approach to the issue of personal post-mortem survival. *Jornal de Estudos Espíritas*. https://doi.org/10.22568/jee.v8.artn.010203

Coelho, H. S. (2012). *Livre-arbítrio e sistema: Conflitos e conciliações em Böhme e Goethe*. https://repositorio.ufjf.br/jspui/handle/ufjf/1696

Cook, E. W., et al. (1983). A review and analysis of "unresolved" cases of the reincarnation type: II. Comparison of features of solved and unsolved cases. *Journal of the American Society for Psychical Research, 77*(2), 115–135.

Cook, E. W., Greyson, B., & Stevenson, I. (1998). Do any near-death experiences provide evidence for the survival of human personality after death? Relevant features and illustrative case reports. *Journal of Scientific Exploration, 12*, 377–406.

DataFolha. (2007). DataFolha: Instituto de pesquisas. http://media.folha.uol.com.br/datafolha/2013/05/02/religiao_03052007.pdf

de Rochas, A. (1911). *Les Vies successives, documents pour l'étude de cette question, Bibliothèque Chacornac, Paris*.

Delorme, A., Beischel, J., Michel, L., Boccuzzi, M., Radin, D., & Mills, P. J. (2013). Electrocortical activity associated with subjective communication with the deceased. *Frontiers in Psychology, 4*. https://doi.org/10.3389/fpsyg.2013.00834

Du Prel, C. (1889). The philosophy of mysticism: Vol. I. George Redway.

Evrard, R. (n.d.). *Institut Métapsychique International | Psi Encyclopedia*. Retrieved August 15, 2021, from https://psi-encyclopedia.spr.ac.uk/articles/institut-m%C3%A9tapsychique-international

Fadiman, J., & Frager, R. (Eds.). (1997). *Essential sufism* (1st ed.). HarperSanFrancisco.

Farnell, L. R. (1921). *Greek hero cults and ideas of immortality: The Gifford lectures delivered in the University of St. Andrews in the year 1920*. Clarendon Press.

Freire, E. S., Rocha, A. C., Tasca, V. S., Marnet, M. M., & Moreira-Almeida, A. (2022). Testing alleged mediumistic writing: An experimental controlled study. *Explore, 18*(1), 82–87. https://doi.org/10.1016/j.explore.2020.08.017

Gallup, G. (2003). *The Gallup poll: Public opinion*. Scholarly Resources.

Gauld, A. (1982). *Mediumship and survival: A century of investigations*. Heinemann.

Geach, P. T. (1979). Truth, love and immortality: An introduction to McTaggart's philosophy.

Greyson, B. (2007). Consistency of near-death experience accounts over two decades: Are reports embellished over time? *Resuscitation, 73*(3), 407–411. https://doi.org/10.1016/j.resuscitation.2006.10.013

Greyson, B. (2010). Seeing dead people not known to have died: "Peak in Darien" experiences: Seeing dead people. *Anthropology and Humanism, 35*(2), 159–171. https://doi.org/10.1111/j.1548-1409.2010.01064.x

Greyson, B. (2015). Western scientific approaches to near-death experiences. *Humanities, 4*(4), 775–796. https://doi.org/10.3390/h4040775

Gurney, E., Myers, F. W. H., & Podmore, F. (1886). *Phantasms of the living* (012058197) (Vol. 1–2). Society for Psychical Research.

Hamann, J. G. (1959). *Hauptschriften erklärt. Bd. II: Sokratische Denkwürdigkeiten. Erklärt von Fritz Blanke*. Gütersloher Verlagshaus Gerd Mohn.

Haraldsson, E. (1995). Personality and abilities of children claiming previous-life memories. *The Journal of Nervous and Mental Disease, 183*(7), 445–451. https://doi.org/10.1097/00005053-199507000-00004

Haraldsson, E. (1997). A psychological comparison between ordinary children and those who claim previous-life memories. *Journal of Scientific Exploration, 1*(3), 323–335.

Haraldsson, E. (2003). Children who speak of past-life experiences: Is there a psychological explanation? *Psychology and Psychotherapy, 76*(Pt 1), 55–67. https://doi.org/10.1348/14760830260569256

Haraldsson, E. (2001). Birthmarks and Claims of Previous-Life Memories: I. The case of Purnima Ekanayake. *Journal of the Society for Psychical Research, 64*(858), 10.

Helck, W., & Otto, E. (1975). *Lexikon der Ägyptologie: Vol. 7 Bd*. Otto Harrassowitz.

Hodgson, R. (1892). A record of observations of certain phenomena of trance. *Proceedings of the Society for Psychical Research, 8*, 1–167.

Hodgson, R. (1897). A further record of observations of certain phenomena of trance. *Proceedings of the Society for Psychical Research, 13*, 284–582.

Holden, J. M., Greyson, B., & James, D. (Eds.). (2009). *The handbook of near-death experiences: Thirty years of investigation*. Praeger Publishers.

Hume, D. (1849). *Essays and treatises on various subjects: With a brief sketch of the author's life and writings*. J.P. Mendum.

Hyslop, J. H. (1901). A further record of observations of certain trance phenomena. *Proceedings of the Society for Psychical Research, 16*, 1–649.

Hyslop, J. H. (1905). *Problem of philosophy or principles of epistemology and metaphysics*. Macmillan.

Inglehart R, & Basanez. (2004). *Human beliefs and values: A cross-cultural sourcebook based on the 1999–2002 values survey*. Delegación Coyoacán, México: Siglo Veintiuno Editores.

James, W. (1886). *Report of the committee on mediumistic phenomena* (1637878). Viking Press.

Johnson, A. (1980). Report on Some Recent Sittings for Physical Phenomena in America. In *Proceedings of the Society For Psychical Research: Vol. LIV–LVI*.

Josephus, F. (1900). *Geschichte des Jüdischen Krieges. Übersetzt und mit Einleitung und Anmerkungen versehen von Dr. Heinrich Clementz, Halle an der Saale (Verlag von Otto Hendel) 1900*. https://opus4.kobv.de/opus4-Fromm/frontdoor/index/index/docId/31965

Kardec, A. (1861). The mediums' book.

Kardec, A. (1890). Posthumous works.

Kean, L. (2017). *Surviving death: A journalist investigates evidence for an afterlife*. Crown Archetype.

Keil, H. H. J., & Stevenson, I. (1999). Do cases of the reincarnation type show similar features over many years? A study of Turkish cases a generation apart. *Journal of Scientific Exploration, 13*(2), 189–198.

Keil, H. H., & Tucker, J. B. (2000). An unusual birthmark case thought to be linked to a person who had previously died. *Psychological Reports, 87*(3 Pt 2), 1067–1074. https://doi.org/10.2466/pr0.2000.87.3f.1067

Keil, H., & Tucker, J. (2005). Children who claim to remember previous lives: Cases with written records made before the previous personality was identified. *Journal of Scientific Exploration, 19*, 91–101.

Kelly, E. W. (2001). Near-death experiences with reports of meeting deceased people. *Death Studies, 25*(3), 229–249. https://doi.org/10.1080/07481180125967

Kleisner, F. (1998). *Körper und Seele bei Georg Christoph Lichtenberg*. Königshausen & Neumann.

Lerner, A. B., Case, J. D., Takahashi, Y., Lee, T. H., & Mori, W. (1958). Isolation of melatonin, the pineal gland factor that lightens melanocytes. *Journal of the American Chemical Society, 80*(10), 2587–2587. https://doi.org/10.1021/ja01543a060

Lodge, O. (1909). *The survival of man*. Wentworth Press.

Lombroso, C. (1909). *After death—What? Spiritistic phenomena and their interpretation*. Small, Maynard & Company.

Lucchetti, G., Daher, J. C., Iandoli, D., Gonçalves, J. P. B., & Lucchetti, A. L. G. (2013). Historical and cultural aspects of the pineal gland: Comparison between the theories provided by Spiritism in the 1940s and the current scientific evidence. *Neuro Endocrinology Letters, 34*(8), 745–755.

MacGregor, G. (1982). *Reincarnation as a Christian hope*. Macmillan.

Mainieri, A. G., Peres, J. F. P., Moreira-Almeida, A., Mathiak, K., Habel, U., & Kohn, N. (2017). Neural correlates of psychotic-like experiences during spiritual-trance state. *Psychiatry Research. Neuroimaging, 266*, 101–107. https://doi.org/10.1016/j.pscychresns.2017.06.006

References

Martial, C., Cassol, H., Antonopoulos, G., Charlier, T., Heros, J., Donneau, A.-F., Charland-Verville, V., & Laureys, S. (2017). Temporality of features in near-death experience narratives. *Frontiers in Human Neuroscience, 11*. https://doi.org/10.3389/fnhum.2017.00311

McTaggart, J. M. E. (1930). *Some dogmas of religion*. Edward Arnold.

Mills, A., Haraldsson, E., & Keil, H. H. J. (1994). Replication studies of cases suggestive of reincarnation by three independent investigators. *Journal of the American Society for Psychical Research, 88*(3), 207–219.

Moore, L. E., & Greyson, B. (2017). Characteristics of memories for near-death experiences. *Consciousness and Cognition, 51*, 116–124. https://doi.org/10.1016/j.concog.2017.03.003

Moraes, L. J., da Barbosa, G. S., de Castro, J. P. G. B., Tucker, J. B., & Moreira-Almeida, A. (2022). Academic studies on claimed past-life memories: A scoping review. *Explore, 18*(3), 371–378 https://doi.org/10.1016/j.explore.2021.05.006

Moreira-Almeida, A. (2008). Allan Kardec and the development of a research program in psychic experiences. *Proceedings of Presented Papers of the Parapsychological Association Convention, 51*, 136–151.

Moreira-Almeida, A. (2012). Research on mediumship and the mind–brain relationship. In *Exploring frontiers of the mind-brain relationship* (pp. 191–213). Springer.

Moreira-Almeida, A., & Santos, F. S. (2012). Exploring frontiers of the mind-brain relationship. *Springer Science & Business Media.* https://doi.org/10.1007/978-1-4614-0647-1

Muldoon, S. J., & Carrington, H. (1984). *The phenomena of astral projection*. Rider.

Myers, F. W. H. (1903). *Human personality and its survival of bodily death* (Vol. 1–2). Longmans, Green and Co.

Nahm, M., Greyson, B., Kelly, E. W., & Haraldsson, E. (2012). Terminal lucidity: A review and a case collection. *Archives of Gerontology and Geriatrics, 55*(1), 138–142. https://doi.org/10.1016/j.archger.2011.06.031

Obst, H. (2009). *Reinkarnation: Weltgeschichte einer Idee*. Verlag C.H. Beck. https://www.chbeck.de/obst-reinkarnation/product/26017

Palmieri, A., Calvo, V., Kleinbub, J. R., Meconi, F., Marangoni, M., Barilaro, P., Broggio, A., Sambin, M., & Sessa, P. (2014). "Reality" of near-death-experience memories: Evidence from a psychodynamic and electrophysiological integrated study. *Frontiers in Human Neuroscience, 8*, 429. https://doi.org/10.3389/fnhum.2014.00429

Paraná, D., Rocha, A. C., Freire, E. S., Lotufo Neto, F., & Moreira-Almeida, A. (2019). An empirical investigation of alleged mediumistic writing: A Case study of Chico Xavier's letters. *The Journal of Nervous and Mental Disease, 207*(6), 497–504. https://doi.org/10.1097/NMD.0000000000000999

Parnia, S. (2007). Do reports of consciousness during cardiac arrest hold the key to discovering the nature of consciousness? *Medical Hypotheses, 69*(4), 933–937. https://doi.org/10.1016/j.mehy.2007.01.076

Pasricha, S. (1996). Phobias in cases of the reincarnation type. *NIMHANS Journal, 14*, 51–55.

Pasricha, S. (2011). Do attitudes of families concerned influence features of children who claim to remember previous lives? *Indian Journal of Psychiatry, 53*(1), 21. https://doi.org/10.4103/0019-5545.75554

Pasricha, S., & Stevenson, I. (1987). Indian cases of the reincarnation type two generations apart. *Journal of the Society for Psychical Research, 54*(809), 239–246.

Perandréa, C. A. (1991). A psicografia à luz da grafoscopia. *Semina, 10*(Ed. Especial), 59–71.

Peres, J. F., Moreira-Almeida, A., Caixeta, L., Leao, F., & Newberg, A. (2012). Neuroimaging during trance state: A contribution to the study of dissociation. *PLoS One, 7*(11), e49360. https://doi.org/10.1371/journal.pone.0049360

Piddington, J. G. (1908). A series of concordant automatisms. In *Proceedings of the Society for Psychical Research: Vol. XXII*.

Playfair, G. L. (2010). *Chico Xavier, a medium of the century*. Roundtable Publishing.

Ring, K., & Cooper, S. (1997). Near-death and out-of-body experiences in the blind: A study of apparent eyeless vision. *Journal of Near-Death Studies, 16*(2), 101–147. https://doi.org/10.1023/A:1025010015662

Rivas, T., Dirven, A., & Smit, R. H. (2016). *The self does not die: Verified paranormal phenomena from near-death experiences*. International Association for Near-Death Studies.

Rocha, A. C. (2001). *A poesia transcendente de Parnaso de alem-tumulo*. Master's dissertation (Universidade Estadual de Campinas [UNICAMP]). Available at http://www.espiritualidades.com.br/Artigos/R_autores/ROCHA_Alexandre_Caroli_tit_Poesia_transcendente_de_Parnaso_de_Alem_Tumulo-A.pdf

Rocha, A. C. (2008). *O caso Humberto de Campos: Autoria literaria e mediunidade*. Doctoral thesis (Universidade Estadual de Campinas [UNICAMP]). Available at https://bv.fapesp.br/pt/dissertacoesteses/74465/o-caso-humberto-de-campos-autoria-literaria-e-mediunidade

Rocha, A. C., Paraná, D., Freire, E. S., Lotufo Neto, F., & Moreira-Almeida, A. (2014). Investigating the fit and accuracy of alleged mediumistic writing: A case study of Chico Xavier's letters. *Explore, 10*(5), 300–308. https://doi.org/10.1016/j.explore.2014.06.002

Sabom, M. B. (1982). *Recollections of Death: A Medical Investigation*. Eweb:40112. https://repository.library.georgetown.edu/handle/10822/792025

Sabom, M. B. (1998). *Light & death: One doctor's fascinating account of near-death experiences*. Zondervan.

Sabom, M. B., & Kreutiziger, S. (1977). Near-death experiences. *The New England Journal of Medicine, 297*(19), 1071. https://doi.org/10.1056/nejm197711102971919

Sánchez-Barceló, E. J., Mediavilla, M. D., Tan, D. X., & Reiter, R. J. (2010). Clinical uses of melatonin: Evaluation of human trials. *Current Medicinal Chemistry, 17*(19), 2070–2095. https://doi.org/10.2174/092986710791233689

Sarraf, M., Woodley of Menie, M. A., & Tressoldi, P. (2021). Anomalous information reception by mediums: A meta-analysis of the scientific evidence. *Explore, 17*(5), 396–402. https://doi.org/10.1016/j.explore.2020.04.002

Schmidt-Leukel, P., & Bauer, E. (Eds.). (1996). *Die Idee der Reinkarnation in Ost und West*. E. Diederichs Verlag.

Scholz, H. (1934). *Goethes Stellung zur Unsterblichkeitsfrage*. Mohr.

Schouten, S. A., & Stevenson, I. (1998). Does the socio-psychological hypothesis explain cases of the reincarnation type? *The Journal of Nervous and Mental Disease, 186*(8), 504–506. https://doi.org/10.1097/00005053-199808000-00011

Severino, P. R. (1992). *A vida triunfa: Pesquisa sobre mensagens que Chico Xavier recebeu*. Editora Jornalística Fé.

Shushan, G. (Ed.). (2009). *Conceptions of the afterlife in early civilizations: Universalism, constructivism, and near-death experience*. Continuum.

Smit, R. H. (2008). Letter to the editor: Further commentary on pam Reynolds's NDE. *Journal of Near-Death Studies, 26*(4), 308–310. https://doi.org/10.17514/JNDS-2008-26-4-p308-310

Souto Maior, M. (2003). *As vidas de Chico Xavier* (2a. ed., rev. e ampliada). Editora Planeta do Brasil.

Stevenson, I. (1961). The evidence for survival from claimed memories of former incarnations the winning essay of the contest in memory of William James. *Postgraduate Medical Journal, 37*(429), 443–444.

Stevenson, I. (1974a). Some questions related to cases of reincarnation type. *Journal of the American Society for Psychical Research, 68*, 396–416.

Stevenson, I. (1974b). *Xenoglossy: A review and report of a case*. University Press of Virginia.

Stevenson, I. (1977a). The explanatory value of the idea of reincarnation. *The Journal of Nervous and Mental Disease, 164*(5), 305–326. https://doi.org/10.1097/00005053-197705000-00002

Stevenson, I. (1977b). Research into the evidence of man's survival after death: A historical and critical survey with a summary of recent developments. *The Journal of Nervous and Mental Disease, 165*(3), 152–170. https://doi.org/10.1097/00005053-197709000-00002

References

Stevenson, I. (1983). American children who claim to remember previous lives. *The Journal of Nervous and Mental Disease, 171*(12), 742–748. https://doi.org/10.1097/00005053-198312000-00006

Stevenson, I. (1990). Phobias in children who claim to remember previous lives. *Journal of Scientific Exploration*, 243–254.

Stevenson, I. (1993). Birthmarks and birth defects corresponding to wounds on deceased persons. *Journal of Scientific Exploration, 7*(4), 403–410.

Stevenson, I. (1997). *Reincarnation and biology: A contribution to the etiology of birthmarks and birth defects*. Praeger.

Stevenson, I. (2000). Unusual play in young children who claim to remember previous lives. *Journal of Scientific Exploration, 14*, 557–570.

Stevenson, I. (2001). *Children who remember previous lives: A question of reincarnation*. McFarland.

Stevenson, I., & Keil, J. (2005). Children of Myanmar who behave like Japanese soldiers: A possible third element in personality. *Journal of Scientific Exploration, 19*, 171–183.

Stevenson, I., & Samararatne, G. (1988). Three new cases of the reincarnation type in Sri Lanka with written records made before verification. *The Journal of Nervous and Mental Disease, 176*(12), 741. https://doi.org/10.1097/00005053-198812000-00008

Tanne, J. H. (2007). Ian Pretyman Stevenson. *BMJ, 334*(7595), 700–700. https://doi.org/10.1136/bmj.39141.529792.65

Tart, C. T. (1968). A psychophysiological study of out-of-the-body experiences in a selected subject. *Journal of the American Society for Psychical Research, 62*, 3–27.

Thonnard, M., Charland-Verville, V., Brédart, S., Dehon, H., Ledoux, D., Laureys, S., & Vanhaudenhuyse, A. (2013). Characteristics of near-death experiences memories as compared to real and imagined events memories. *PLoS One, 8*(3), e57620. https://doi.org/10.1371/journal.pone.0057620

Trent-Von Haesler, N., & Beauregard, M. (2013). Experiências de quase morte em parada cardíaca: Implicações para o conceito de mente não local. *Archives of Clinical Psychiatry, 40*, 197–202. https://doi.org/10.1590/S0101-60832013000500005

Tucker, J. B. (2006). *Life before life: A scientific investigation of children's memories of previous lives*. Piatkus.

Tucker, J. B. (2008). Children's reports of past-life memories: A review. *Explore, 4*(4), 244–248. https://doi.org/10.1016/j.explore.2008.04.001

van Lommel, P. (2011). Near-death experiences: The experience of the self as real and not as an illusion: Near-death experiences and the self. *Annals of the New York Academy of Sciences, 1234*(1), 19–28. https://doi.org/10.1111/j.1749-6632.2011.06080.x

van Lommel, P. (2013). Non-local consciousness a concept based on scientific research on near-death experiences during cardiac arrest. *Journal of Consciousness Studies, 20*(1–2), 1–2.

van Lommel, P., van Wees, R., Meyers, V., & Elfferich, I. (2001). Near-death experience in survivors of cardiac arrest: A prospective study in the Netherlands. *The Lancet, 358*(9298), 2039–2045. https://doi.org/10.1016/S0140-6736(01)07100-8

Weaver, Z. (n.d.). *Our History | spr.ac.uk*. Retrieved August 15, 2021, from https://www.spr.ac.uk/about/our-history

6
The Weight of the Whole Body of Evidence for Life After Death

Based on the evidence presented on Chap. 5, it is easy to see why these spiritual experiences (SE) have tended to generate beliefs in some sort of dualism (in the sense that mind is something beyond the body) and survival, in virtually any culture throughout the ages (Bozzano, 1997; Lang, 1898; Lombroso, 1909; Shushan, 2009; Tylor, 1871). Moreover, it is remarkable that contemporary scientific evidence leads to conclusions very similar to those which have been perennially advocated by most religions, spiritual practices, and philosophies throughout history. The universal beliefs in an afterlife—assumed to be "primitive" by radical secularists—proved consistent in the face of scientifically controlled investigation (Kripal, 2010).

At first glance, the evidence, indeed, strongly suggests that human beings have something beyond their bodies and this something can separate itself from the body and even survive death. It is necessary, however, to analyze if this first impression survives a closer and more rigorous examination of the evidence in light of the main alternative hypotheses. As we have discussed at Chap. 5, there are four main hypotheses to account for the facts. Two of them are "conventional"—(1) deliberate fraud and (2) misinterpretation of chance or of experiences generated by neurophysiology or unconscious mind activity—and two go beyond the conventional physicalist perspective: (3) living agent psi (LAP) (also called ESP—extrasensory perception) and (4) survival.

In dealing with a complex topic such as the nature of human consciousness, it is essential to avoid simplistic approaches and not be naïve enough to think that one single hypothesis can explain all the evidence previously discussed. Indeed, we will now argue that the conventional explanations and LAP can explain much of the evidence, but neither of them in isolation or even in conjunction can explain all the evidence. Our point here is that survival of consciousness is a necessary hypothesis, in conjunction with the other three, to make sense of all the available evidence.

We will take here a phenomenological approach to survival of consciousness, and not delving into the ultimate nature of what survives or the mechanisms by which it works (Chibeni & Moreira-Almeida, 2007). Based on this empirical or phenomenological approach, we will propose a "pragmatic dualism" as a working hypothesis, i.e., human beings have two aspects: physical body and something beyond that survives bodily death (the thinking being, soul, spirit, consciousness, mind, etc.). However, as yet, we are unable to define whether we are ultimately composed of only one substance (monism), two substances (dualism), or more (Hyslop, 1901). This is more a metaphysical than an empirical question (Chibeni, 2020; Moreira-Almeida, 2012). In addition, the mechanisms by which these two aspects of human beings interact are still unclear. But, as discussed at section "There Is No Mechanism for How the Mind Would Influence the Brain", several theories have been proposed to explain this interaction. These, along with new ones, should be developed and tested.

Considering we are still in a pre-paradigmatic period regarding the nature of consciousness and the mind-brain relationship, theoretical pluralism is better than theoretical monism (Chibeni & Moreira-Almeida, 2007; Moreira-Almeida & Araujo, 2015). It is desirable that a wide variety of paradigm candidates should emerge and that they should be tested based on the broadest possible range of empirical evidence. As stated by the philosopher of science Imre Lakatos, there should be an open-minded Darwinian competition of research programs (paradigm candidates), so the fittest will improve and prevail in the "natural selection" of rigorous tests based on logic and empirical evidence (Lakatos & Musgrave, 1970).

When discussing the scientific evidence for survival, as well as for any other scientific hypothesis, it is critical to bear in mind that there is no such thing as a crucial experiment or definitive single proof (Chalmers, 1982; Popper, 1963). For example, a recent paper discussing how to discover the causes of diseases reminds us that "no single study can provide a definitive answer to the questions of causality" (Ohlsson & Kendler, 2020, p. 637), because any study and research method has its own strengths and limitations.

Based on this, key epidemiologists propose that the evidence is much stronger when similar findings result from different studies, using different methods and performed by different researchers—this is often called *triangulation* (Ohlsson & Kendler, 2020; VanderWeele, 2021). Sir Austin Bradford Hill[1] proposed the now classic nine criteria for determining if an association is causal. One of them is triangulation, where consistent findings have been "repeatedly observed by different persons, in different places, circumstances and times … the same answer has been reached in quite a wide variety of situations and techniques" (Hill, 1965, p. 296–297). Discussing his criteria, he arrived at a conclusion that applies very well to our overall assessment of the empirical evidence for survival:

> What they can do, with greater or less strength, is to help us to make up our minds on the fundamental question - is there any other way of explaining the set of facts before us, is there any other answer equally, or more, likely than cause and effect? (Hill, 1965, p. 299)

In addition, with regard to the search for strong evidence and weighing the pros and cons to reach a scientific conclusion:

---

[1] A pivotal epidemiologist in the process of incorporating statistics and randomization into medicine and in proving that smoking causes lung cancer (Wilkinson, 1997).

In asking for very strong evidence I would, however, repeat emphatically that this does not imply crossing every 't', and swords with every critic, before we act. All scientific work is incomplete. (Hill, 1965, p. 300)[2]

Now, coming back to survival, as we have seen, around 150 years of survival research has produced a body of evidence that is larger and stronger than most people are aware of or have even imagined being possible to attain.[3]

Hypotheses Alternative to Survival

Chapter 5 discussed each of the main lines of evidence: mediumship, OBE/NDE, and CORT. It is time now to weigh the whole body of evidence. We will briefly discuss the three main alternatives to the survival theory that would explain the available evidence.

Fraud

Undoubtedly, there have been many fraud cases, but there are many good studies with positive results that have properly controlled for this possibility. For example, Hodgson, an expert in unmasking fraud, investigated the medium Mrs. Piper. Sitters unknown to her were introduced anonymously, she was taken to England and kept under continuous surveillance, and detectives were hired to investigate if she tried to get previous information about the sitters and deceased. In decades of investigation, no reasonable charge of fraud was raised against her (Braude, 2003; Gauld, 1982). In addition, a recent meta-analysis of controlled, triple-blind studies with mediums also found positive results (Sarraf et al., 2021). Fraud could also not account for the similarities in many features in cases studied in different times and places. For example, it is not credible that fraudulent parents all over the world (many of them illiterate and with no access to publications on the topic) have colluded to produce similar cases of children claiming memories of previous lives, such as the age they start and finish talking about these memories, frequent claims of violent death, the rarity of claimed deceased famous personalities, birthmarks, etc. (Matlock, 2019).

---

[2] A similar balance between naïve hasty conclusions and a cynical denial of unideal pieces of evidence was recently emphasized in epidemiology (Ohlsson & Kendler, 2020).

[3] As stated by Hans Eysenck, one of the most influential psychologists of the twentieth century, the available evidence for survival after death "is capable of scientific consideration and is of very high quality" (Eysenck & Sargent, 1993, p. 151).

Chance, Cryptomnesia, Fabrications of the Unconscious Mind, and Other Conventional Sources

These are certainly a major source, perhaps *the* major one, for alleged empirical evidence for survival of consciousness claimed by laypeople. However, the control and exclusion of these conventional sources are the main target (in addition to control for fraud) of any minimally qualified researcher in this field. As discussed at length in Chap. 5, there are plenty of well-conducted studies whose results cannot be reasonably explained by these usual suspects. This is especially the case when the pieces of information provided were unknown to the witnesses (e.g., parents or sitters) and to the person who produces the evidence (e.g., medium, children, or NDE experiencer). Under controlled situations, the information is often too abundant, precise, private, and specific to be accounted for by chance or previous knowledge obtained through ordinary means by mediums or children claiming previous lives. In addition to information, it is also necessary to account for unlearned skills, birthmarks, personality traits, mannerisms, etc.

Living Agent Psi (LAP)

These two previous usual suspects have been consistently addressed since the dawn of survival research and were basically considered insufficient to account for all the available, empirical evidence by the vast majority of researchers who have studied it in-depth. They usually agree on the need for some non-conventional explanation, typically LAP and/or survival (Almeder, 1992; Bem, 2005; Braude, 2003; Eysenck & Sargent, 1993; Gauld, 1982; James, 1909; Lombroso, 1909; Richet & De Brath, 1923; Sudduth, 2016). Of course, as one might expect, there are also some who think that everything could be explained by the usual suspects (Lester, 2005; Martin & Augustine, 2015). However, they often slip into several of the epistemological pitfalls discussed in Chap. 3 and elsewhere (Moreira-Almeida, 2006).

It is astonishing to realize that the major challenge for the survival hypothesis is not a conventional explanation but a "paranormal" one, LAP. Hundreds of studies have established, on good controlled empirical grounds, the reality of the so-called parapsychological (psi) phenomena. A recent review of the large body of experimental evidence on psi was published in a major journal of the American Psychological Association and concluded that:

> The evidence provides cumulative support for the reality of psi, which cannot be readily explained away by the quality of the studies, fraud, selective reporting, experimental or analytical incompetence, or other frequent criticisms. The evidence for psi is comparable to that for established phenomena in psychology and other disciplines. (Cardeña, 2018, p. 663)

The primary question is if the empirical evidence for survival discussed in this essay can properly be explained by LAP, i.e., telepathy and clairvoyance, rendering the survival hypothesis superfluous. We argue that this is not the case. In addition, it is worth noting that many authors, from Allan Kardec (1860) to Ian Stevenson (1977), via Carl du Prel (1939/1899), Cesare Lombroso (1909), Ernesto Bozzano (1995/1938), J. B. Rhine, and William McDougall along the way, have argued (and we agree with them) that LAP, by itself, suggests a transcendent/non-physical/spiritual aspect of human beings that might, in principle, survive bodily death (Alvarado, 2012; Griffin, 1997). LAP is a major challenge to conventional physicalist views of mind. It is hard to figure out how electrical and chemical exchanges between neurons could explain thought transference between people and visual perception of distant objects.[4]

Regardless of the implications of LAP for the ultimate nature of the human mind, we will list below the main reasons that make LAP an unlikely explanation for the whole body of evidence and leaves survival of consciousness as the most logical and parsimonious hypothesis. Much evidence for survival, to be explained by LAP, would require assuming the existence of a level of LAP much higher than we have evidence for, for example, unlearned skills (e.g., xenoglossy and poetry), birthmarks, and consistent production of a series of personal veridical information (unknown to anyone present) (Bozzano, 1995; Kelly, 2013; Lombroso, 1909).

Because these alleged psi skills are well beyond anything previously documented to exist, it is often called *super-psi*. Some authors (Braude, 2003; Sudduth, 2009) have claimed that this is not a reasonable objection, because we actually don't know the nature of psi and, hence, we also don't know its limits. We agree that the nature of psi is not yet understood; however, we can test its possibilities and limits by analyzing spontaneous and experimental cases. For example, what have we been able to document about the range of telepathic interactions among the living? Sure, we may find new telepathic possibilities beyond that which has previously been recorded. However, it would be very suspicious if this alleged new kind of telepathy (hypothesized to explain the evidence for survival) is only detected under circumstances indicative of survival such as in a mediumistic trance or among children claiming to have lived previous lives. In order to make the case for a stronger sort of LAP (instead of evidence for survival), such alleged new telepathic capacity should also be detected in cases not related to survival.

Some authors (Braude, 2003; Sudduth, 2009, 2021) argue that LAP and survival would require the same level of psi functioning to be postulated to explain the best evidence of survival. This would be the case because the explanation for survival requires some sort of psi. For example, mediumship would actually be a telepathic interaction between the medium and a discarnate personality, instead of between living personalities, as in LAP. This is a good point, and parity indeed exists if we are comparing the telepathic interaction between the medium and a sitter or a discarnate personality. However, mediumship would require a much more complex LAP if the medium produces a series of information, personality traits, and skills not possessed or known by the sitters and not available from any single source (apart

---

[4] The incompatibility between LAP and strict physicalist views of human nature is also acknowledged by most so-called skeptics and is a major reason often raised by them to justify an a priori denial of any LAP claim (Reber & Alcock, 2020).

from the deceased personality). In order to make the case for a LAP-related explanation, it would require the medium to be in telepathic contact with several other living persons in different and distant places and have a clairvoyant perception of different locations to be able to collect the specific pieces of information (and only those) that were known by the alleged deceased personality. In other words, this would require almost an omniscient capacity that would involve a psychic connection with people and places that are located far from the mediums and with which they have no previous contact or emotional rapport, conditions known to enhance LAP (Bozzano, 1995; Carpenter, 2012). In addition, the medium would have to put together these widely scattered pieces of information and, at the same time, express them in a coherent way through the impersonation of the deceased, expressing them as coming from him/her. It seems much more complex than inferring that the medium is in telepathic contact with just one mind, i.e., the communicating deceased personality. This is one example where clearly there is no parity on the postulated level of psi functioning.

This previous objection to LAP as an explanation for the best evidence of survival becomes even stronger when incorporating what Braude (2003) calls *crippling complexity*. The main idea is that psi (like all human abilities) is highly situation-sensitive, having a causal network that may be impaired by many different factors. For example, the performance of a star soccer player in a given match depends on a wide variety of factors: physical fitness, training, emotional balance, payment, desire to collaborate with teammates or to shine alone, crowd reaction, ball and pitch conditions, etc. Then:

> The more dense and extensive that network is, the more obstacles any particular psychic inquiry or effort must navigate in order to succeed ... the more potentially wide-ranging and virtuosic we take psi to be, the less likely it becomes that a person's psi could produce an extended and accurate trance persona, or provide all the detailed, intimate information found in the most astonishing survival cases—and even more so, to do these things consistently. (Braude, 2021, p. 159–160)

Based on this, Braude argues that crippling complexity favors survival over LAP as it is more parsimonious, because the "survivalist scenario requires no more than the integrity of a single causal connection between the psychic subject and a postmortem individual" while "super-psi requires multiple sources of information" each of them with "distinct causal connection" (Braude, 2003, p. 305) subject to obstacles related to the *crippling complexity*.

The survival hypothesis also benefits from the fact that, in many instances, the person who produces evidence of survival does not demonstrate evidence of high levels of psi functioning in other situations or domains of life. For example, many (if not most) children in reincarnation cases, reporting consistent and accurate memories as well as fears, interests, and personality traits related to a deceased personality, do not show any evidence of telepathy or clairvoyance in their lives. So, in order to explain these cases simply in terms of LAP, it would be necessary to postulate that these children have high levels of telepathy and clairvoyance even if we have no evidence of this, except for the evidence related to the alleged previous life. It might be possible, but it does not seem very credible.

In addition, the medium and the children would also need to experience and present all this information purportedly obtained by LAP as if coming from a specific deceased person, who they usually don't know. Since the anomalous experiences discussed here have been widespread across time and space, this LAP hypothesis would also require a perennial and universal unconscious willingness to deceive (the experiencers themselves and others) that would lead them to impersonate, very skillfully, specific deceased persons. Finally, it would also require a high level of dramatic capability in order to perform the "dramatic play of personality" in such a way as to convince the deceased's close relatives and acquaintances (Hyslop, 1901).

It is also worth noting that the available evidence shows that all of a personality that we would expect to survive actually survives (Hyslop, 1901), i.e., memory, affections, skills, motivations, personal way of dealing with things, etc. In fact, several survival researchers have found more than they had initially expected. For example, Ian Stevenson (2007) initially only expected to find claims of verbal memories allegedly related to previous lives. When he actually started to investigate CORT, he also found several other features of the previous personality: skills, personality traits, birthmarks, phobias, etc. Similarly, Erlendur Haraldsson (Haraldsson, 2012a), quite unexpectedly, found that CORT children often have symptoms of posttraumatic stress disorder, despite having no clear major traumatic experience in this life. However, these symptoms make sense in light of their alleged memories of past-life traumatic events.

In summary, the hypothesis of LAP (without survival) for the best evidence of survival would require us to assume, *simultaneously*:

1. The existence of a level of LAP much higher than we have evidence for from experimental or spontaneous cases.
2. The person being, at the same time, in telepathic and clairvoyant contact with several different (and often unknown) living persons and distant places to be able to collect the specific pieces of information (and only those) that were known by the alleged deceased personality.
3. Being able to navigate through the crippling complexity of these required multiple sources of information, each of them with a complex causal network subject to many obstacles.
4. The person who produces evidence of survival often does not show evidence of high levels of LAP in other situations or domains of life.
5. Being able to integrate into a coherent whole all these isolated pieces of information collected from different and distant sources.
6. Perennial and universal unconscious willingness to deceive that would lead them to impersonate, very skillfully, specific deceased persons often unknown to them.
7. This impersonation would be so competent that it would convince close acquaintances and include a wide range of personal features that were not initially expected, even by experts in the field.

Each of these assumptions is improbable in isolation. This unlikelihood becomes even more pronounced if we need to multiply these low probabilities in order to make them work together and fully explain the many good pieces of evidence for survival presented in this essay.

In addition to the major obstacles described above for the usual suspects or LAP explaining away each specific good piece of evidence for survival, the problem becomes even more challenging when we consider the convergence of the whole range of different kinds of evidence. The full body of evidence provides a compelling case for triangulation—converging evidence (from different researchers, from a broad variety of research groups, countries, and methods for a wide range of phenomena investigated), which mutually reinforce each other, pointing to the same conclusion: survival of consciousness. For each piece of evidence, taken in isolation, we might potentially concede an unlikely explanation of undetected fraud, chance, cryptomnesia, or LAP; however, when we need to multiply dozens of times the concession to these unlikely possibilities, it becomes less and less credible. This would be the case in trying to explain away altogether the whole body of evidence produced by Leonora Piper and Chico Xavier or the multiple high-quality cases of reincarnation, NDE, and OBE.[5] The compelling strength of this convergence of evidence was classically emphasized by Frederic Myers (1903) and Ernesto Bozzano (1995/1938) and has also been bolstered by contemporary researchers (Almeder, 1992; Braude, 2003; Gauld, 1982; Griffin, 1997; Stevenson, 1977) (Fig. 6.1).

Furthermore, it is worth taking into consideration the explanatory value of the survival hypothesis, to explain not only the wide range of survival evidence *per se* but also a wide range of challenging phenomena that defy full explanation by

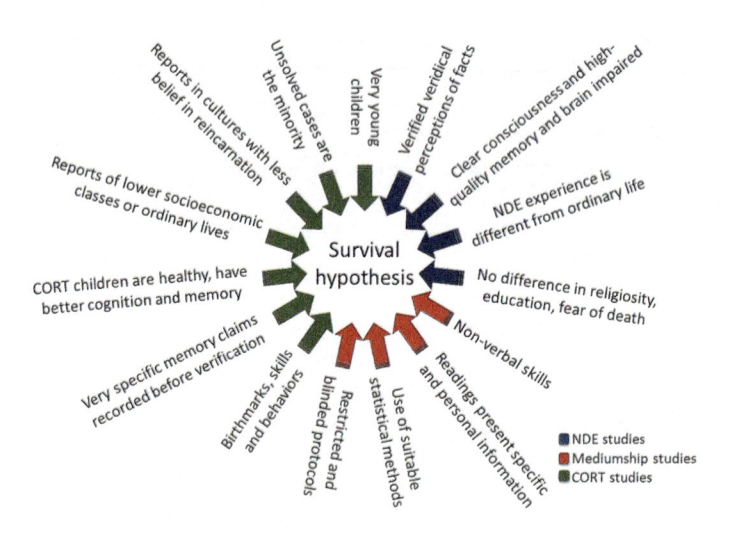

Fig. 6.1 The compelling strength of the convergence of evidence for survival hypothesis

---

[5] As stated by James Hyslop, Professor of Philosophy at Columbia University, after in-depth personal investigation of Mrs. Piper: "My preference for the spiritistic theory after facing the problems just indicated rests on a very simple basis, and it is that I am not prepared to build any altars to Mrs. Piper's brain" (Hyslop, 1901, ch. I and VI).

physicalist accounts of human beings that attempt to pigeonhole all human experience only in terms of biology and environmental influences. If, in addition to these two explanatory factors, we add a mind that survives the body (in close bidirectional interaction with it during bodily life), a wide range of apparently anomalous phenomena (from a physicalist perspective) would make sense, e.g., placebo effect, child prodigies, artistic inspiration, gender identity issues, children's behavioral and psychological characteristics and differences between identical twins not explained by genetics or the environment, etc. (Kelly, 2007; Myers, 1903; Stevenson, 2000).

The survival hypothesis has a series of virtues desirable in a good scientific hypothesis: empirical adequacy (its empirical consequences are true), broadness of scope (explains a wide range of phenomena), prediction of new kinds of phenomena (such as PTSD symptoms in CORT), and simplicity (Chibeni & Moreira-Almeida, 2007). Sure, the evidence is not perfect; it never is. As discussed previously, there is no such thing as definitive proof, and it is always possible to raise more than one hypothesis to explain a given scientific observation. Scientific revolutions such as Heliocentrism and Darwinism took place not because they were perfect theories. In truth, they were incomplete, with some inconsistent empirical findings and with no clear explanatory mechanism. They prevailed because they possessed the epistemological virtues discussed above, especially the support of a broad spectrum of empirical findings and being able to explain with simplicity a wide range of natural phenomena (Ghiselin, 1969; Kuhn, 1970; Principe, 2011; Shapin, 1996).

In a similar vein, the hypothesis that human beings have a consciousness that is not produced by the body and survives its death is supported by a wide range of different kinds of converging empirical evidence. In addition, it has a stronger explanatory power connecting and making sense of a large and broad array of experiences, either anomalous experiences (e.g., mediumship, NDE, OBE, ELE, CORT) or "common" experiences (e.g., volition, freewill, *qualia* experience and the unity of consciousness, etc.). There is no competing paradigm candidate that is equally well supported by empirical evidence and capable of accounting for all the facts together. This has been the conclusion reached by many (probably most) highly qualified, scientific and philosophical minds (from diverse intellectual and geographical backgrounds) who painstakingly performed and published comprehensive analyses of the available evidence for survival[6]:

- UK: biologist and co-discoverer of the theory of evolution by natural selection Alfred Russel Wallace (1866), psychical researcher Frederic Myers (1903), physicists Sir Oliver Lodge (1929) and Sir William Crookes (1972), and psychologists Alan Gauld (1982) and Hans Eysenck and Sargent (1993).
- USA: philosophers James Hyslop (1901), Robert Almeder (1992), David Griffin (1997), and Stephen Braude (2003), psychiatrist Ian Stevenson (Kelly, 2013;

---

[6] This list can be much enlarged with the vast majority of the 43 authors of the 29 winning essays of the essay contest on the evidence for survival promoted by BICS (Bigelow Institute for Consciousness Studies) in 2021. All the wining essays are available for download at https://www.bigelowinstitute.org/contest_winners3.php

Stevenson, 1977), and psychologists Edward and Emily Kelly (2007) and Charles Tart (2009).
- France: scholar Allan Kardec (Moreira-Almeida, 2008), astronomer Camille Flammarion (1921), physician Gustav Geley (1919, 1925), and probably the Nobel laureate in Medicine, Charles Richet (Bubb, 1936; Richet, 1997).
- Italy: psychical researcher Ernesto Bozzano (1995) and psychiatrist and founder of criminal anthropology, Cesare Lombroso (1909).
- Germany: philosopher Carl du Prel (1889) and astrophysicist Johann Zöllner and Massey (1880).
- Iceland: psychologist Erlendur Haraldsson (2012b; Haraldsson & Matlock, 2017).
- Russia: psychical researcher Alexander Aksakof (1991)
- Australia: psychical researcher Richard Hodgson (1897).
- Brazil: engineer Hernani Guimarães Andrade (1983), psychiatrist Alexander Moreira-Almeida, and neuropsychiatrist Franklin Santos (Moreira-Almeida & Santos, 2012).
- Portugal: physician and businessman Luis Portela (2021).

References

Aksakof, A. (1991). *Animismo e espiritismo*. Federação Espírita Brasileira.
Almeder, R. F. (1992). *Death and personal survival: The evidence for life after death*. Rowman & Littlefield Publishers.
Alvarado, C. S. (2012). Psychic phenomena and the mind-body problem: Historical notes on a neglected conceptual tradition. In *Exploring frontiers of the mind-brain relationship* (pp. 35–51). https://doi.org/10.1007/978-1-4614-0647-1_3
Andrade, H. G. (1983). *Morte, renascimento, evolução*. Pensamento.
Bem, D. J. (2005). Review of the afterlife experiments. *Journal of Parapsychology, 69*, 173–183.
Bozzano, E. (1995). *Animismo ou Espiritismo? Qual dos dois explica o conjunto dos fatos? [Animism or spiritism? Which of the two expalins the ensemble of facts?]*. Federação Espírita Brasileira.
Bozzano, E. (1997). *Povos primitivos e manifestações supranormais*. Editora Jornalistica FE.
Braude, S. E. (2003). *Immortal remains: The evidence for life after death*. Rowman & Littlefield.
Braude, S. E. (2021). Matlock's theoretical offenses. *Journal of Scientific Exploration, 35*(1), 155–165. https://doi.org/10.31275/20212086
Bubb, M. (1936, May 30). Richet accepted survival before he "died." *Psychic News, (n. 210)*, 7.
Cardeña, E. (2018). The experimental evidence for parapsychological phenomena: A review. *American Psychologist, 73*(5), 663–677. https://doi.org/10.1037/amp0000236
Carpenter, J. C. (2012). *First sight: ESP and parapsychology in everyday life* (pp. xii, 487). Rowman & Littlefield.
Chalmers, A. F. (1982). *What is this thing called science? An assessment of the nature and status of science and its methods* (2nd ed.). University of Queensland Press.
Chibeni, S. (2020). Spiritism: An experimental approach to the issue of personal post-mortem survival. *Jornal de Estudos Espíritas*. https://doi.org/10.22568/jee.v8.artn.010203
Chibeni, S. S., & Moreira-Almeida, A. (2007). Remarks on the scientific exploration of "anomalous" psychiatric phenomena. *Archives of Clinical Psychiatry, 34*, 8–16. https://doi.org/10.1590/S0101-60832007000700003
Crookes, W. (1972). *Crookes and the spirit world*. Souvenir.

References

Du Prel, C. (1889). *The philosophy of mysticism: Vol. I*. George Redway.
Eysenck, H. J., & Sargent, C. (1993). *Explaining the unexplained: Mysteries of the paranormal*. Prion.
Flammarion, C. (1921). *Death and its mystery (3 vol)*. Century.
Gauld, A. (1982). *Mediumship and survival: A century of investigations*. Heinemann.
Geley, G. (1919). *De l'Inconscient au Conscient*. Felix Alcan.
Geley, G. (1925). Essai de Revue générale Et d'interprétation du Spiritisme. B.P.S.
Ghiselin, M. T. (1969). *The triumph of the Darwinian method*. University of California Press.
Griffin, D. R. (1997). *Parapsychology, philosophy, and spirituality: A postmodern exploration*. State University of New York Press.
Haraldsson, E. (2012a). Cases of the reincarnation type and the mind–brain relationship. In A. Moreira-Almeida & F. Santana Santos (Eds.), *Exploring Frontiers of the mind-brain relationship* (pp. 215–231). Springer. https://doi.org/10.1007/978-1-4614-0647-1_11
Haraldsson, E. (2012b). *The departed among the living: An investigative study of afterlife encounters*. White Crow Books.
Haraldsson, E., & Matlock, J. G. (2017). *I saw a light and came here: Children's experiences of reincarnation*. White Crow Books.
Hill, A. B. (1965). The environment and disease: Association or causation? *Proceedings of the Royal Society of Medicine, 58*(5), 295–300. https://doi.org/10.1177/003591576505800503
Hodgson, R. (1897). A further record of observations of certain phenomena of trance. *Proceedings of the Society for Psychical Research, 13*, 284–582.
Hyslop, J. H. (1901). A further record of observations of certain trance phenomena. *Proceedings of the Society for Psychical Research, 16*, 1–649.
James, W. (1909). The final impressions of a psychical researcher. In G. Murphy & R. O. Ballou (Eds.), *William James on psychical research* (pp. 309–325). Viking.
Kelly, E. F. (2007). *Irreducible mind: Toward a psychology for the 21st century*. Rowman & Littlefield.
Kelly, E. W. (2013). In E. W. Kelly (Ed.), *Science, the self, and survival after death: Selected writings of Ian Stevenson*. Rowman & Littlefield Publishers. https://rowman.com/ISBN/9781442221147/Science-the-Self-and-Survival-after-Death-Selected-Writings-of-Ian-Stevenson
Kripal, J. J. (2010). *Authors of the impossible: The paranormal and the sacred*. The University of Chicago Press.
Kuhn, T. S. (1970). *The structure of scientific revolutions*; 2nd ed., enlarged). University of Chicago Press.
Lakatos, I., & Musgrave, A. (1970). *Criticism and the growth of knowledge*. Cambridge University Press.
Lang, A. (1898). *The making of religion*. Longmans.
Lester, D. (2005). *Is there life after death? An examination of the empirical evidence*. McFarland. http://www.loc.gov/catdir/toc/ecip056/2004029515.html
Lodge, O. (1929). *Why I believe in personal immortality*. (5th ed.). Cassell and company, ltd.
Lombroso, C. (1909). *After death—What? Spiritistic phenomena and their interpretation*.
Martin, M., & Augustine, K. (2015). *The myth of an afterlife: The case against life after death*. Rowman & Littlefield.
Matlock, J. G. (2019). *Signs of reincarnation: Exploring beliefs, cases, and theory*. Rowman & Littlefield.
Moreira-Almeida, A. (2006). Book review—Is there life after death? An examination of the empirical evidence—David Lester. *Journal of Near-Death Studies, 24*(4), 245–254. https://doi.org/10.17514/JNDS-2006-24-4-p245-254
Moreira-Almeida, A. (2008). *Allan Kardec and the development of a research program in psychic experiences* (Vol. 51, pp. 136–151).
Moreira-Almeida, A. (2012). Research on mediumship and the mind–brain relationship. In *Exploring frontiers of the mind-brain relationship* (pp. 191–213). Springer.
Moreira-Almeida, A., & Araujo, S. F. (2015). Does the brain produce the mind? A survey of psychiatrists' opinions *Archives of Clinical Psychiatry, 42*(3), 74–75.

Moreira-Almeida, A., & Santos, F. S. (2012). Exploring frontiers of the mind-brain relationship. *Springer*.https://doi.org/10.1007/978-1-4614-0647-1

Myers, F. W. H. (1903). *Human personality and its survival of bodily death* (Vol. 1–2). Longmans, Green and co.

Ohlsson, H., & Kendler, K. S. (2020). Applying causal inference methods in psychiatric epidemiology: A review. *JAMA Psychiatry, 77*(6), 637–644. https://doi.org/10.1001/jamapsychiatry.2019.3758

Popper, K. R. (1963). *Conjectures and refutations: The growth of scientific knowledge*. Routledge & Kegan Paul.

Portela, L. (2021). *The science of spirit: Parapsychology, enlightenment and evolution.*

Principe, L. (2011). *The scientific revolution: A very short introduction*. Oxford University Press.

Reber, A. S., & Alcock, J. E. (2020). Searching for the impossible: Parapsychology's elusive quest. *The American Psychologist, 75*(3), 391–399. https://doi.org/10.1037/amp0000486

Richet, C. (1997). *No limiar do mistério*. Ideba.

Richet, C., & De Brath, S. (1923). *Thirty years of psychical research: Being a treatise on metapsychics*. W. Collins Sons.

Sarraf, M., Woodley of Menie, M. A., & Tressoldi, P. (2021). Anomalous information reception by mediums: A meta-analysis of the scientific evidence. *Explore, 17*(5), 396–402. https://doi.org/10.1016/j.explore.2020.04.002

Shapin, S. (1996). *The scientific revolution*. University of Chicago Press.

Shushan, G. (Ed.). (2009). *Conceptions of the afterlife in early civilizations: Universalism, constructivism, and near-death experience*. Continuum.

Stevenson, I. (1977). Research into the evidence of man's survival after death: A historical and critical survey with a summary of recent developments. *The Journal of Nervous and Mental Disease, 165*(3), 152–170. https://doi.org/10.1097/00005053-197709000-00002

Stevenson, I. (2000). The phenomenon of claimed memories of previous lives: possible interpretations and importance. *Med Hypotheses, 54*(4), 652–659. https://doi.org/10.1054/mehy.1999.0920

Stevenson, I. (2007). Half a career with the paranormal. *Archives of Clinical Psychiatry, 34*(1), 150–155. https://doi.org/10.1590/S0101-60832007000700019

Sudduth, M. (2009). Super-psi and the survivalist interpretation of mediumship. *Journal of Scientific Exploration, 23*(2), 167–193.

Sudduth, M. (2016). *A philosophical critique of empirical arguments for post-mortem survival*. Palgrave Macmillan.

Sudduth, M. (2021). Book review of signs of reincarnation: Exploring beliefs, cases, and theory by James Matlock. *Journal of Scientific Exploration, 35*(1), 183–208. https://doi.org/10.31275/2021208

Tart, C. T. (2009). *The end of materialism: How evidence of the paranormal is bringing science and spirit together*. Noetic Books New Harbinger Publications.

Tylor, E. B. (1871). *Primitive culture: Researches into the development of mythology, philosophy, religion, art, and custom*. J. Murray.

VanderWeele, T. J. (2021). Can sophisticated study designs with regression analyses of observational data provide causal inferences? *JAMA Psychiatry, 78*(3), 244–246. https://doi.org/10.1001/jamapsychiatry.2020.2588

Wallace, A. R. (1866). *The scientific aspect of the supernatural: Indicating the desirableness of an experimental enquiry by men of science into the alleged powers of clairvoyants and mediums*. http://books.google.com/books?vid=BL:A0018283198Google_Books.

Wilkinson, L. (1997). Sir Austin Bradford Hill: Medical statistics and the quantitative approach to prevention of disease. *Addiction, 92*(6), 657–666.

Zöllner, J. C. F., & Massey, C. C. (1880). *Transcendental physics, an account of experimental investigations: From the Scientific treatises* (Translated by C.C. Massey).

7
Cultural Barriers to a Fair Examination of the Available Evidence for Survival

Survival is the simplest and most straightforward explanation for the whole diversified body of evidence. However, this is often refuted because of previous metaphysical (often dogmatic) commitment to physicalism (Martin & Augustine, 2015) or, in some cases, to religious dogmas (Coelho, 2022). Usually, if there was not an a priori emotional or intellectual obligation to physicalism (Lycan, 2009), it would be hard not to see all this evidence as pointing to the survival of consciousness.

Although pervasive and overly repeated, the statement that science is materialistic and rejects the idea of a surviving soul is a myth. As we mentioned in Chap. 3, pervasive physicalism in the academic environment is an anomaly of the last century. Contemporary physicalists often claim they are taking a "skeptical" approach, but frequently different from a healthy and true critical skepticism, which means a kind of agnosticism, a suspension of belief (including the physicalist one) until consistent rational and empirical evidence is obtained. Contrary to this original definition, a "dogmatic skepticism" (which is an oxymoron) arose, characterized by radical and irreversible denial of some ideas because of an unshakable commitment to another. One who rejects discussion of alternative perspectives and potentially falsifying evidence is not a skeptic but a dogmatic denier, usually because of dogmatic devotedness to another belief, usually physicalism.

This sort of pseudoskepticism, an a priori denial that something different from the subject's beliefs may be true, was also reported by William James who invited eight Harvard intellectuals (Sommer, 2013, p. 219):

> to come to my house at their own time, and sit with a medium [Mrs. Piper] for whom the evidence already published in our *Proceedings* had been the most noteworthy. Although it means at worst the waste of the hour for each, five of them decline the adventure. ... one of the five colleagues who declined my invitation is widely quoted as an effective critic of our evidence. So runs the world away!

James also tells of another psychologist[1] who also refused to study Mrs. Piper, replying it would be "useless, for if he should get such results as I report, he would (being suggestible) simply believe himself hallucinated." The most telling thing is

---

[1] Subsequently identified as Hugo Münsterberg, "the most extensively covered US psychologist in the New York Times, in 1909 he seized the opportunity to 'expose' the famous physical medium, Eusapia Palladino" (Sommer, 2013, p. 273)

that this colleague "writes *ex cathedra* on the subject of psychical research, declaring that there is nothing in it" (Sommer, 2013, p. 216).[2]

The main barrier to a truly rational and scientific analysis of survival research is materialist scientism, expressed as: "science needs or has proved materialism and all educated persons know that"; "it has been proved that brain produces mind"; "evidence of survival must be false, because we know that only matter exists." But these assertions only beg the question, since it is precisely the physicalism that is in question.

Interestingly, religious dogmatism may add much to materialist scientism in terms of opposition to this research into the survival of human personality (e.g., denial of evidence of reincarnation or mediumship among many Christians).[3]

And it is always important to remember how blatant opposition to ideas is intimidating when scholars are afraid to investigate survival because of a lack of funding and risk of jeopardizing their careers and reputations. Two telling examples of the impact of peer pressure as major obstacles to psychical research are the cases of Richet (1898) and Freud (Josephson, 2017).

The assumption of materialist philosophy as a natural part of the contemporary scientific paradigms has too many historical/philosophical roots to trace in the present essay, but most of us can relate it to positivism and other forms of reductionist scientism.

A classic example is the first philosophy of Ludwig Wittgenstein. In his *Tractatus Logico-Philosophicus*, he wrote that: "1- The world is everything that is the case"; "1.1 - The world is the totality of facts, not of things"; "2.1 - We make images of these facts to ourselves" (Wittgenstein, 1960). This naïve and simplistic objectivism was used as a weapon by those interested in eliminating spiritualist or dualist concepts as "meaningless," but, because words do not correspond exactly to facts, this exercise in cleansing failed in its task of building a new "scientific language" from scratch. All facts are interpreted, as Wittgenstein himself would admit in the second phase of his philosophical work.

In his *Philosophical Investigations*, Wittgenstein argued that meaning depends on broader semantic agreements that work as language games (Wittgenstein, 1958, §539), helping to pave the way to a more social and cultural concept of knowledge than the straightforward—and wrong—conception of the positivists. As positivism and naïve scientism imploded in their own terms, the philosophical support for physicalism was ultimately damaged to its metaphysical core.

---

[2] A more recent example was a paper entitled "There is nothing paranormal about near-death experiences," published in a prestigious neuroscience journal claiming that all reports of NDE could be explained by conventional neuroscience (Mobbs & Watt, 2020). However, one of the authors acknowledged that they avoided looking at evidence of OBE with veridical perception in NDE. In summary, they purposely excluded the main paranormal evidence related to NDE and concluded that there is nothing paranormal about NDE! (Greyson et al., 2020; Tsakiris, 2012).

[3] However, the Anglican theologian MacGregor (1982) argues persuasively, in historical, theological, and biblical terms, that reincarnation is not only compatible with Christianity but actually "could be an enrichment of the Christian hope" (MacGregor, 1982, p. ix).

An extremely contrasting philosophical project, such as the one by G. W. F. Hegel, came to conclusions very close to those of Wittgenstein. Hegel created the possibility of a sociological understanding by showing precisely how many of our concepts about reality and objects are intersubjective (Hegel, 1970). Contrary to the previous philosophical and scientific notion that we perceive reality and simply get it right or wrong, Hegel consistently argued that most of our perceptions, judgments, and thoughts are heavily dependent on language, culture, religion, and art, in short, cultural history. The fact that two very different and very influential philosophical traditions ended up with very similar conclusions is significant for the overall state of our contemporary understanding of reason and scientific thought.

We can summarize this conclusion, viz., contemporary concept of experience and evidence changed significantly from the positivist and mechanist myth of direct perception of reality. We now know that all perception and what we consider evidence, although objective, are heavily dependent on intersubjective/social concepts and criteria worth considering in a more robust concept of experience/evidence. To show conclusive evidence, therefore, involves deep social understanding of cultural trends that define conclusive evidence, and without addressing these trends, the search for ("pure") evidence is fruitless.

As stated by the historian and philosopher of science Thomas Kuhn:

> The transfer of allegiance from paradigm to paradigm is a conversion experience that cannot be forced. Lifelong resistance, particularly from those whose productive careers have committed them to an older tradition of normal science
> ... [A] generation is sometimes required to effect the change ... Though some scientists, particularly the older and more experienced ones, may resist indefinitely, most of them can be reached in one way or another. (Kuhn, 1970, p. 151–152)

The psychological process of forming and changing beliefs is deeply complex and not straightforward, especially regarding beliefs with higher potential impact on a person's worldview, values, ethics, and philosophy of life. In addition, several researchers have emphasized how much the personal witnessing of phenomena related to survival impacted their acceptance of the survival hypothesis, more than any evidence obtained indirectly by reading scientific reports (Bozzano, 1995; Lombroso, 1909). As stated by the journalist Leslie Kean in the conclusion to her recent book on scientific evidence for the afterlife:

> For me, such convincing studies and case accounts have not had the same impact as have my personal experiences, even though these experiences are likely to be less convincing to anyone else. (Kean, 2017, p. 359)

As we are aware of the cultural blind spot for the given mass of evidence, what can be done? Firstly, stimulate intellectual humility about what we know—and do not know—and about the universe and a truly scientific spirit that combines open-mindedness and rigorousness. So, as the challenge is enormous, intense cultural efforts await willing, committed actors to, among others, deconstruct the many prevalent myths, ideologies, and prejudices about the subject in academic and general audiences; present the best evidence to academic and general audiences; and stimulate the development of new methods and theories (candidates for paradigm)

(Gauld, 1982; Moreira-Almeida & Araujo, 2017). Finally, the few groups that do swim against the tide to shed light on the overwhelming evidence in favor of survival need to be fostered and multiplied. Small and committed groups or even individuals can do much, as demonstrated by the examples of Ian Stevenson (Kelly, 2013) and the SPR (Gauld, 1968, 1982, p. 266).

Spiritual phenomena such as those analyzed in this book cry out for serious investigation as well as openness and courage to accept the radical implications they entail for our understanding of nature and of ourselves. Given the persuasiveness of the evidence, there is no reason for us to hide away in this self-imposed obscurity any more.

References

Bozzano, E. (1995). *Animismo ou Espiritismo? Qual dos dois explica o conjunto dos fatos?* *[Animism or spiritism? Which of the two explains the ensemble of facts?]*. Federação Espírita Brasileira.
Coelho, H. S. (2022). *História da Liberdade Religiosa: da Reforma ao Iluminismo*. Vozes.
Gauld, A. (1968). *The founders of psychical research*. Schocken Books.
Gauld, A. (1982). *Mediumship and survival: A century of investigations*. Heinemann.
Greyson, B., Holden, J. M., & van Lommel, P. (2020). 'There is nothing paranormal about near-death experiences' revisited: Comment on Mobbs and Watt. *Trends in Cognitive Sciences, 16*(9), 445.
Hegel, G. W. F. (1970). *Werke in 20 Bänden*. Ed. Moldenhauer, Eva; Michel, Karl M. Suhrkamp.
Josephson-Storm, J. A. (2017). *The myth of disenchantment: Magic, modernity, and the birth of the human sciences*. The University of Chicago Press.
Kean, L. (2017). *Surviving death: A journalist investigates evidence for an afterlife*. Crown Archetype.
Kelly, E. W. (2013). *Science, the self, and survival after death: Selected writings of Ian Stevenson* (E. W. Kelly, Ed.). Rowman & Littlefield Publishers.
Kuhn, T. S. (1970). *The structure of scientific revolutions* (2nd ed., enlarged.). University of Chicago Press.
Lombroso, C. (1909). *After death—What? Spiritistic phenomena and their interpretation*.
Lycan, W. G. (2009). Giving dualism its due. *Australasian Journal of Philosophy, 87*(4), 551–563. https://doi.org/10.1080/00048400802340642
MacGregor, G. (1982). *Reincarnation as a Christian hope*. Macmillan.
Martin, M., & Augustine, K. (2015). *The myth of an afterlife: The case against life after death*. Rowman & Littlefield.
Mobbs, D., & Watt, C. (2020). There is nothing paranormal about near-death experiences: How neuroscience can explain seeing bright lights, meeting the dead, or being convinced you are one of them. *Trends in Cognitive Sciences, 15*(10), 447–449.
Moreira-Almeida, A., & Araujo, S. F. (2017). The mind-brain problem in psychiatry: Why theoretical pluralism is better than theoretical monism. *Dialogues in Philosophy, Mental and Neuro Sciences, 10*, 23–25.
Richet, C. (1898). On the conditions of certainty. *Proceedings of the Society for Psychical Research, 14*, 152–157.
Sommer, A. (2013). Crossing the boundaries of mind and body. Psychical research and the origins of modern psychology. In *Science and Technology Studies: Vol. PhD*. University College London.

Tsakiris, A. (2012, March 20). Dr. Caroline Watt defends, there is nothing paranormal about near-death experiences. *Skeptiko - Science at the Tipping Point*. https://skeptiko.com/165-dr-caroline-watt-defends-there-is-nothing-paranormal-about-near-death-experiences/

Wittgenstein, L. (1958). *Philosophical investigations*. Macmillan.

Wittgenstein, L. (1960). *Tractatus Logico-Philosophicus*. Routledge & Kegan Paul.

8
Conclusion

When analyzing the whole body of survival evidence with a rational, rigorous, and open mind, it is hard to resist the conclusion that *Survival of Human Consciousness after Permanent Bodily Death* is a fact of nature. Indeed, this has consistently been the conclusion reached by the vast majority of humanity, including those who performed careful philosophical and scientific analysis thereof. The best evidence for survival is the compelling convergence of findings from dozens of highly qualified scientists in a wide range of anomalous experiences (mediumship, apparitions, ELE, OBE, NDE, CORT, etc.) that mutually reinforce each other because they point to the same conclusion: survival of consciousness. To reject survivalist explanations for those findings would require postulating, simultaneously, a sequence of very unlikely assumptions and facts. Survival is the most simple, comprehensive, and natural explanation for the empirical data discussed in this essay.

The rejection of the conclusion for survival by many scholars is a historical exception concentrated mainly in the twentieth century. It seems that this rejection of evidence for survival is strongly related to the cultural prejudices associated with the physicalist mindset of materialist scientism. Based on the misguided (and often dogmatic) assumption that physicalism is a proven fact and necessary for rational and scientific analysis (Coelho, 2012; Moreira-Almeida et al., 2018), there has been an a priori rejection of the possibility of survival as a hypothesis and a dogmatic denial of the supporting empirical evidence. The consequence is that many (perhaps most) scholars and even average educated laypersons are not aware of the compelling empirical evidence for survival.

In order to overcome these barriers to a fair analysis of the evidence for survival, two major steps are needed:

- Deconstruct misguided physicalist and anti-spiritual philosophical, historical, and methodological assumptions that cloud a fair consideration and analysis of the survival hypothesis and the supporting empirical evidence.

- Fairly present the whole body of evidence indicative of survival of consciousness.

These steps are the two main goals of the present essay. To advance a rigorous academic discussion on consciousness survival, in addition to a wider dissemination of what is already known, it is crucial to support the development and a Darwinian-like competition of research programs (Moreira-Almeida & Araujo, 2017). These research programs would not only develop paradigm candidates to account for the whole body of evidence but also test them under a diversity of high-quality, empirical studies with different, but complementary, methodological approaches (Chibeni & Moreira-Almeida, 2007; Moreira-Almeida & Lotufo-Neto, 2017).

References

Chibeni, S. S., & Moreira-Almeida, A. (2007). Remarks on the scientific exploration of "anomalous" psychiatric phenomena. *Archives of Clinical Psychiatry (São Paulo), 34*, 8–16. https://doi.org/10.1590/S0101-60832007000700003
Coelho, H. S. (2012). *Livre-arbítrio e sistema: Conflitos e conciliações em Böhme e Goethe.* https://repositorio.ufjf.br/jspui/handle/ufjf/1696
Moreira-Almeida, A., & Araujo, S. d. F. (2017). The mind-brain problem in psychiatry: Why theoretical pluralism is better than theoretical monism. *Dialogues in Philosophy, Mental and Neuro Sciences, 10*, 23–25.
Moreira-Almeida, A., & Lotufo-Neto, F. (2017). Methodological guidelines to investigate altered states of consciousness and anomalous experiences. *International Review of Psychiatry, 29*(3), 283–292. https://doi.org/10.1080/09540261.2017.1285555
Moreira-Almeida, A., Araujo, S. F., & Cloninger, C. R. (2018). The presentation of the mind-brain problem in leading psychiatry journals. *Brazil Journal of Psychiatry, 40*(3), 335–342. https://doi.org/10.1590/1516-4446-2017-2342

Index

A
Academic myth, 9
Afterlife belief, 1
Alleged neural mechanisms, 15
Alleged personality skills, 30
American Psychological Association, 64
American scientific psychology, 34
Ancient Egyptians, 33
Anecdotal evidence, 34
Anecdotal references, 6
Animal Magnetism, 34
Anomalies, 17
Antiquity, 7, 27, 47
Anti-scientific, 9
a priori disqualification, 9
Aristotle, 8
Automatic writing, 41

B
Babylonian *epic Gilgamesh*, 6
Belief in divinatory, 6
Beyond Physicalism (Book), 18
Biological immortality, 6
Birthmarks, 64
Birthmarks and birth defects, 52
Birthmarks/birth defects, 51
Blind individuals, 44

C
Cardiac arrest, 46
Cartesian dualism, 21
Cases of the reincarnation type (CORTs), 46, 48–53, 67

Chaffin, 39
Chico's literary mediumistic production, 38
Children make statements, 53
Clairvoyance, 64
Clairvoyant perception, 66
Communication, 28, 33
Communicators, 35, 36
Complementary, methodological approaches, 82
Consciousness, 30, 40, 42, 43, 45, 46, 61, 62, 64, 65, 68, 69
Consciousness activities, 16
Consciousness survival, 82
Contemporary concept, 77
Contemporary scientific paradigms, 76
Continuity of character and memory, 29
Contradictory philosophical consequences, 19
Convenient recognition devices, 29
Conventional explanations, 20
Conventional sources, 64
Crash cart, 44
Crippling complexity, 66
Cross-correspondence, 35, 36
Cryptomnesia, 64, 68
Cultural prejudices, 81

D
Darwinian-like competition, 82
Death
    life after Life after death
Destitute sick, 37
Discarnate personalities, 34–35
Disruptive scientific discoveries, 18
DNA testing, 29

© The Author(s), under exclusive license to Springer Nature Switzerland AG 2022
A. Moreira-Almeida et al., *Science of Life After Death*, SpringerBriefs in Psychology, https://doi.org/10.1007/978-3-031-06056-4

Dogmatic denial, 17
Dogmatic devotedness, 75
Dogmatic positions, 5
Dogmatic skepticism, 27, 75
Dogmatism, 3
Dramatic play of personality, 67
Dreams, 7
Drop-in communications, 37
Drug use, 45
Dualism, 21

E
Electrical stimulation, 15
Electroencephalogram (EEG), 41
Elementary education, 36
Emotional fragility, 41
Empirical adequacy, 69
Empirical evidence, 2
    dogmatic denial, 17
    hypothesis, 27
    preponderance, 17
    priori rejection of survival, 22
Ethics, 8
Evidence, 1
    mediumship Mediumship
    NDEs/OBEs, 42–46
    reincarnation Reincarnation
    survival, 38
    survival after death Survival after death
Ex cathedra, 76
Existential arguments, 9
Existential courage, 9
Explanatory gap, 16
Extrasensory perception (ESP), 61

F
Firewood (charcoal), 52
Fortuitous occurrence, 8
Fraud, 61, 63, 64, 68
Fraud hunter, 34

G
George Pelham (GP), 35

H
Hallucination, 45
Humanity, 46, 53
Human philosophical understanding, 5
Human skill, 36
Hypoxic/confusional states, 46

I
Immortality of the soul, 1
    empirical proof, 6
    idea of survival, 5
    metaphysical arguments, 10
    sequitur, 8
    spiritual world, 8
Indirect evidence, 29
Influential philosophical traditions, 77
Intellectual humility, 27, 77
Interactionist dualism, 21
Internet age, 39
Intersubjective/social concepts, 77
Intrinsic mechanisms, 18

J
JP's circumstances of death, 37

K
Katharine, 35

L
Law of classical physics, 18
Life after death, 1
    evidence Evidence
Living agent Psi (LAP)
    accurate memories, 66
    affections, 67
    anomalous experiences, 67
    Australia, 70
    Brazil, 70
    clairvoyance, 64
    common experiences, 69
    consciousness, 69
    CORTs, 67
    crippling complexity, 66
    dramatic play of personality, 67
    evidence, 64
    experimental evidence, 64
    explanation, 66
    France, 70
    Germany, 70
    hypothesis, 67
    Iceland, 70
    implications, 65
    Italy, 70
    memory, 67
    motivations, 67
    non-conventional explanation, 64
    omniscient capacity, 66
    past-life traumatic events, 67

personalities, 65
Portugal, 70
Russia, 70
scientific revolutions, 69
skills, 65, 67
super-psi, 65
survival hypothesis, 68, 69
survival research, 64
telepathy, 64, 65
transcendent/non-physical/spiritual aspect, 65
triangulation, 68
UK, 69
USA, 69

M
Mannerisms, 64
Materialist metaphysics, 28
Materialist philosophy, 76
Materialist scientism, 76
Meaningless, 76
Medical conditions, 45
Mediumistic writing, 35
Mediumship
  Ancient Egyptians, 33
  anecdotal evidence, 34
  ASPR, 34
  blind protocols, 40
  Chaffin, 39
  challenging and complex cases, 42
  coincidence, 41
  communication, 33
  emotional fragility, 41
  fraud and superstition, 41
  IMI, 34
  mediumistic phenomena, 39
  meta-analysis, 40
  Palladino, Eusapia, 39
  Piper, Leonora, 34–36
  proxy sitter, 40
  qualitative investigations, 40
  religious and materialist dogmatism, 33
  research protocols, 41
  SPECT, 40
  Spiritism, 34
  spiritual and physical realm, 33
  SPR, 34, 41
  Xavier, Chico, 36–39
Memory, 30
Mental and psychological properties, 29
Metaphysical rejection, 9
Methodological Exclusion of the Transcendent, 28
Mind, 29, 41, 48, 61, 62, 65, 66, 69
Mind-body dualism, 43
Mind-brain relationship, 62
Misguided philosophical reasoning, 19
Modern Spiritualism, 34
Muslims, 47
Myanmar, 50

N
Naïve and simplistic objectivism, 76
Nature, 20
Near-death and out-of-body experiences (NDEs/OBEs)
  communication, 45
  concept, 43
  consciousness, 45
  cultural factors, 43
  descriptions, 42
  features, 46
  hypoxic/confusional states, 46
  life-threatening events, 42
  person's values and beliefs, 45
  psychological/cognitive processes, 45
  survival, 45
  terminal lucidity, 43
  unusual/extraordinary, 42
  veridical perceptions, 43–45
Neurocentrism, 16
Neurohormonal dysregulation, 45
Neurophysiology, 61
Neuroscience
  associated changes, 13
  dualist conclusion, 15
  evidence, 13
  founders, 14
  in vivo electrical stimulation, 14
  non-physicalist views of mind, 13
  physicalist view of mind, 13
Newton's law of gravitation, 18
Nihilism, 9
Non-mental substance, 9
Non-physicalist views, 19

O
Oarapsychological (psi) phenomena, 64
Occam's razor, 16
Omniscient capacity, 66
Open-minded Darwinian competition, 62
Open-mindedness, 77
Opposition to ideas, 76
Optimism, 9
Oxygen deficiency, 45

P

Palladino, Eusapia, 39
Panpsychism, 21
Parnassus from Beyond the Tomb (book), 38
Past lives, 48, 49, 52
Penfield's homunculus, 14
Perceived doctrine, 47
Perfect certainty, 6
Person's life, 53
Personal identity, 2, 28
Personality traits, 30, 64
Persuasiveness, 78
Phallic tomb monuments, 6
Philosophical Investigations, 76
Philosophical project, 77
Phobias, 51
Physicalism, 10
    academic environment, 75
    metaphysical assumption, 19
    perennial, 19
    and spiritualism, 19
Pineal gland/epiphysis, 38
Piper, Leonora, 34–36
Plato's approach, 8
Plausibility, 17
Posttraumatic stress disorder, 50, 67
Pragmatic dualism, 62
Prejudice, 2, 3
Pre-paradigmatic period, 19
Principle of Parsimony, 16
Proceedings, 75
Promissory materialism, 16
Proof, 62, 69
Proxy sitter, 40
Pseudonyms, 34
Pseudoskepticism, 75
Psychic hormones, 38
Psychical Research, 34
Psychography, 37
Psychological problems, 53
Psychological process, 77
Psychophysiological nature, 28
PTSD symptoms in CORT, 69

Q

Quantum interactive dualism, 18

R

Radical implications, 78
Rational and scientific analysis, 81
Rational appreciation, 5

Rational thinking
    structure, 8
    tools, 7
Recognition device, 29
Reincarnation
    antiquity, 47
    birthmarks and birth defects, 52
    Christianity, 47
    concept, 47
    CORTs, 46, 48–53
    cultures, 52
    features, 50
    historical and archaeological research, 46
    humanity, 53
    hypothesis, 53
    idea, 46
    immortality, 47
    Islam, 47
    Lichtenberg, Georg, 48
    in modern-day western societies, 47
    Muslims, 47
    perceived doctrine, 47
    philosophical argument, 47
    psychological problems, 53
    researchers, 50
    solved and unsolved cases, 53
    stresses, 48
    survival, 52
Religious and materialist dogmatism, 33
Religious dogmas, 75, 76
Religious felling, 9
Revolutionary discoveries, 18
Rhetorical strategies, 15
Rigorousness, 77

S

Sacred transcendent, 5
Scholars, 51
Science, 19, 27
Science *vs.* materialism, 9
Scientific achievements humanity, 20
Scientific discoveries, 17
Scientific knowledge, 16
Scientific truth, 3
Second World War, 45
Secularization myth, 1
Single photon emission computed tomography (SPECT), 40
Skeptics, 65
Social constructions, 28
Society for Psychical Research (SPR), 20
Sociological understanding, 77

Soul, 33, 34, 47, 62
Soul/mind/spirit after death, 5
Spirit, 62
Spiritism, 34
Spiritual experiences (SE), 7, 27, 28, 61
Spiritual practices, 61
Spiritual reality, 7
Spiritual traditions, 5
SPR members, 34, 41
Straightforward—and wrong—conception, 76
Stylistic features, 38
Subjectivity, 8
Substance dualism, 21
Supernaturalism, 21
Super-psi, 65
Survival, 38, 39, 42, 43, 45, 47, 48, 52
    assumptions against, 3
    barrier analysis, 81
    compelling convergence, 81
    evidence, 28, 75
    historical exception, 81
    human personality, 1
    preliminary objections, 22
    scientific and rational concept, 6
    scientific evidence, 3
    SPR, 20
Survival after death, 1, 2, 61–63
    chance, 64
    consciousness, 61
    conventional sources, 64
    cryptomnesia, 64
    fraud, 63
    human consciousness, 61
    LAP Living agent Psi (LAP) mechanisms, 62
    pragmatic dualism, 62
    pre-paradigmatic period, 62
    scientific evidence, 62
    triangulation, 62
    unconscious mind, 64
Survival hypothesis, 14, 77
Survival of consciousness, 2, 5, 81

Survival of Human Consciousness after Permanent Bodily Death, 81
Survival research, 2, 3, 20
Swedish medium Emanuel Swedenborg, 8

T
Telepathic and clairvoyant contact, 67
Telepathy, 64, 65
Terminal lucidity, 43
Theoretical mechanism, 17
Transcendental philosophy, 8
Transfer of allegiance, 77
Transmissive organ, 15
Transmissive theories, 15
Triangulation, 62, 68
Triple-blind studies, 63

U
Unconscious mind, 61, 64
Universality, 5
Unlearned skills, 64, 65
Unrestricted certainty, 27
Unsolved cases, 53

V
Veridical perceptions, NDEs/OBEs, 43–45
Visionary dreamlike states, 6
Visual perception, 65
Visual representations, 44

X
Xavier, Chico, 36–39
Xenoglossy, 39

Z
Zoroastrianism, 7

Printed in Dunstable, United Kingdom